Workplace Safety

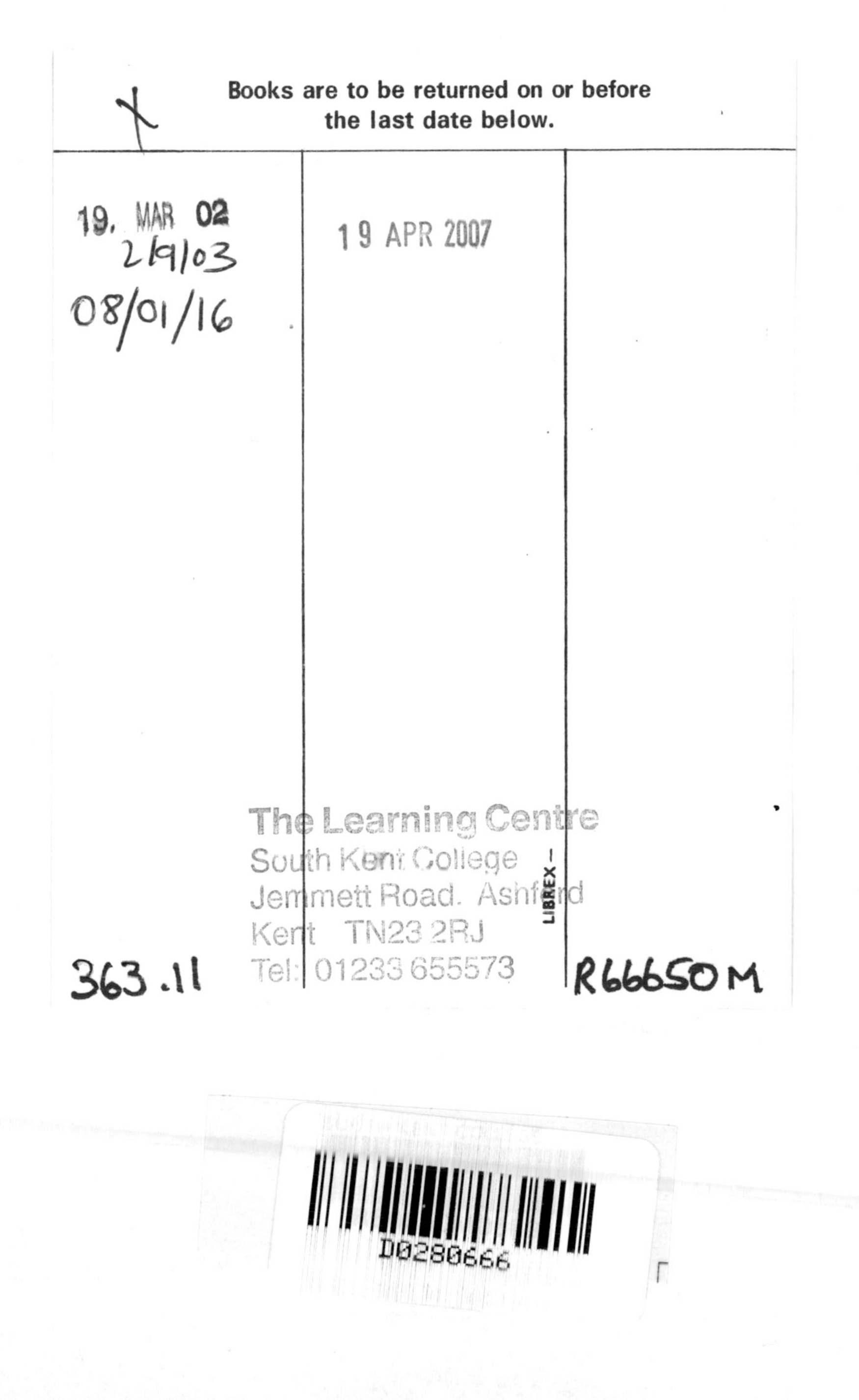

Safety at Work Series*

Volume 1 – Safety Law

Volume 2 – Risk Management

Volume 3 – Occupational Health and Hygiene

Volume 4 – Workplace Safety

---

*These four volumes are available as a single volume, *Safety at Work*, 5th edition.

Workplace Safety

Volume 4 of the Safety at Work Series

Edited by
John Ridley and John Channing

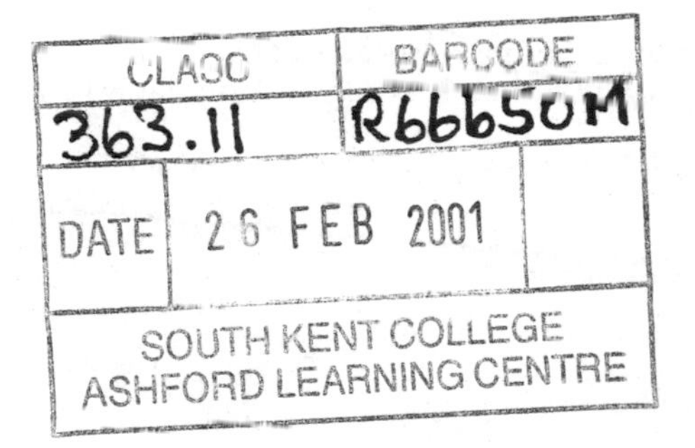

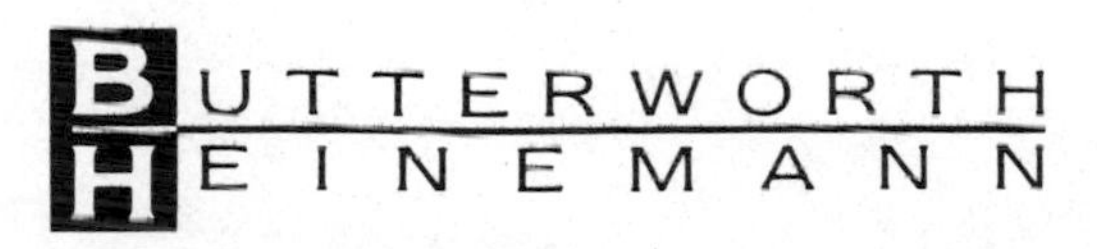

OXFORD AUCKLAND BOSTON JOHANNESBURG MELBOURNE NEW DELHI

Butterworth-Heinemann
Linacre House, Jordan Hill, Oxford OX2 8DP
225 Wildwood Avenue, Woburn, MA 01801-2041
A division of Reed Educational and Professional Publishing Ltd

A member of the Reed Elsevier plc group

First published 1999

British Library Cataloguing in Publication Data
A catalogue record for this book is available from the British Library

Library of Congress Cataloguing in Publication Data
A catalogue record for this book is available from the Library of Congress

ISBN 0 7506 4560 1

Composition by Genesis Typesetting, Laser Quay, Rochester, Kent
Printed and bound in Great Britain by
Biddles Ltd, Guildford and King's Lynn

Contents

Foreword

Frank J. Davies CBE, O St J, *Chairman, Health and Safety Commission*

My forty years experience of working in industry have taught me the importance of health and safety. Even so, since becoming Chairman of the Health and safety Commission (HSC) in October 1993, I have learned more about the extent to which health and safety issues impact upon so much of our economic activity. The humanitarian arguments for health and safety should be enough, but if they are not the economic ones are unanswerable now that health and safety costs British industry between £4 billion and £9 billion a year. Industry cannot afford to overlook these factors and needs to find a way of managing health and safety for its workers and for its businesses.

In his foreword to the third edition of this book my predecessor, Sir John Cullen, commented on the increasing impact of Europe in the field of health and safety, most notably through European Community Directives. We have since found this to be very much so. I believe that the key challenge health and safety now faces is to engage and influence the huge variety of businesses, particularly small businesses, and to help them manage health and safety more effectively. I would add that the public sectors, our largest employers these days, also should look at their management of health and safety to ensure they are doing enough.

Many businesses are willing to meet their legal obligations if given a gentle prompt and the right advice and HSC is very conscious of the importance of having good regulations which are practicable and achievable.

It is, of course, vital and inescapable that an issue as critical as health and safety should be grounded in sound and effective legislation.

This book covers many of these and other important health and safety developments, including environmental and industrial relations law which touch on this area to varying degrees. I welcome the contribution it makes towards the goal of reaching and maintaining effective health and safety policies and practices throughout the workplace.

Preface

Health and safety is not a subject in its own right but is an integration of knowledge and information from a wide spectrum of disciplines. *Safety at Work* reflects this in the range of chapters written by experts and in bringing the benefits of their specialised experiences and knowledge together in a single volume.

While there is a continuing demand for a single volume, many managers and safety practitioners enter the field of safety with some qualifications already gained in an earlier part of their career. Their need is to add to their store of knowledge specific information in a particular sector. Equally, new students of the subject may embark on a course of modular study spread over several years, studying one module at a time. Thus there appears to be a need for each part of *Safety at Work* to be available as a stand-alone volume.

We have met this need by making each part of *Safety at Work* into a separate volume whilst, at the same time, maintaining the cohesion of the complete work. This has required a revision of the presentation of the text and we have introduced a pagination system that is equally suitable for four separate volumes and for a single comprehensive tome. The numbering of pages, figures and tables has been designed so as to be identified with the particular volume but will, when the separate volumes are placed together as a single entity, provide a coherent pagination system.

Each volume, in addition to its contents list and list of contributors, has appendices that contain reference information to all four volumes. Thus the reader will not only have access to the detailed content of the particular volume but also information that will refer him to, and give him an overview of, the wider fields of health and safety that are covered in the other three volumes.

In this way we hope we have kept in perspective the fact that while each volume is a separate part, it is only one part, albeit a vital part, of a much wider spectrum of disciplines that go to make occupational health and safety.

John Ridley
John Channing
October 1998

Contributors

John Adamson, LRSC, FBIOH, Dip. Occ. Hyg., MIOSH, RSP
Manager Health, Safety, Hygiene and Fire, Kodak Chemical Manufacturing
Chris Buck, BSc(Eng), MIEE, FIOSH, RSP
Consultant
Ray Chalklen, MIFireE
Fire and Security Manager, a major pharmaceutical manufacturer
John Channing, MSc(Chem), MSc(Safety), FIOSH, RSP
Manager Health, Safety and Environment, Kodak Manufacturing
Edwin Hooper
R. Hudson, FIOSH, RSP, FRSH, ASSE
Construction safety consultant
John McMullen, BSc, CEng, MIMechE
Eagle Star Insurance Co. Ltd
John Ridley, BSc(Eng), CEng, MIMechE, FIOSH, DMS
Professor P. Waterhouse, PhD, BSc, FIOSH, FIRM

Introduction

Much of the work undertaken by safety advisers requires an understanding of technical industrial processes. Even in a single factory unit the safety adviser may be called upon to advise on avoiding the hazards from a chemical reaction, guarding particular types of machinery, the standards of safe working to be expected of a building contractor, the precautions to be taken to prevent fire and the fire fighting equipment that should be provided, etc.

To carry out his duties effectively, the safety adviser should have an understanding of basic physics and chemistry and of the current safety techniques for reducing the risks associated with the more commonly met industrial processes. This book considers some of these processes and the basic sciences from which they stem.

Chapter 1

Science in engineering safety

J. R. Ridley

1.1 Introduction

In the construction of machines, plant and products, materials are selected because they have particular physical and chemical properties. Wood, metals, concrete, plastics and other substances all have their uses but there are limitations as to what they can do and how long they can do it. Properties may change with use, temperature, operating atmosphere, contamination by surrounding chemicals and for many other reasons. It is necessary to know the properties of the materials and how and why they have been used so that an assessment can be made of whether likely changes in the properties may give rise to hazards.

These properties stem from the chemical and physical characteristics of the different materials and substances used and their behaviour under certain conditions can determine the safety or otherwise of a process or operation. This chapter looks at some of the characteristics and properties of materials in common use, their application, circumstances of use and possible causes of hazards.

1.2 Structure of matter

Everything that we use in our work and daily life is made up of chemical substances, by themselves or in combination of one sort or another. Each substance consists of elements which are the smallest part of matter that can exist by itself. In its free state, an element comprises one or more atoms. When atoms combine together they form molecules of the element or, if different atoms combine, of compounds. The ratio in which atoms combine is determined by their combining power or valency.

Atoms are made up of three particles:

- protons which have a unit mass and carry a positive charge,
- neutrons which have a unit mass but carry no charge, and
- electrons which have negligible mass (i.e. 1/2000 proton) but carry a negative charge.

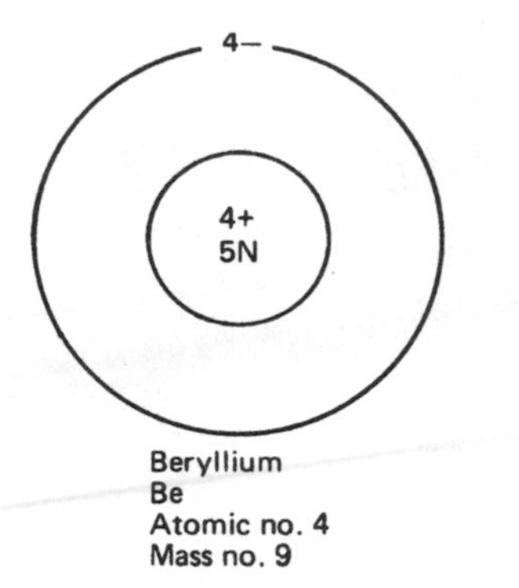

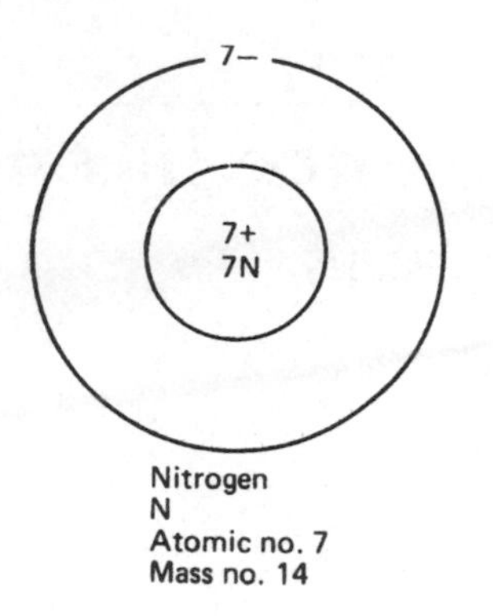

Figure 1.1 Atomic structures

The core or nucleus of the atom consists of protons and neutrons with electrons travelling in orbits around the nucleus (*Figure 1.1*). Elements normally have no overall charge since the number of protons is matched by an equal number of electrons. However, it is possible to upset this balance by removing either a proton or an electron resulting in the atom carrying a charge when it is said to be ionised.

In chemistry, atoms are given 'atomic numbers' which equal the number of protons or electrons in the atom. They are also given 'mass numbers' which equal the sum of the number of protons plus neutrons. The mass number is always equal to or greater than 2 × the atomic number except in the case of hydrogen. Some elements can occur in conditions where they have the same atomic number, and hence the same name, but different mass numbers; they are then known as isotopes and are generally referred to as nuclides. Very large heavy atoms, such as uranium, can be unstable and easily break down to produce smaller atoms with the production of particles or energy. These atoms are radioactive and provide the source of energy in nuclear reactors.

Approximately 100 different atoms have been identified and each has been given a name and a coded symbol which usually is the first one or two letters of its name: carbon – C, lithium – Li, titanium – Ti, etc. Exceptions in this coding system arise because when it was evolved in the early 1800s some chemicals were still known by their Latin names, such as copper (cuprum – Cu) and tin (stannum – Sn).

When atoms join together their molecular formulae are written as groups of atomic symbols to indicate the number of those atoms present to form a stable molecule.

Molecular formulae

Br_2	bromine
O_2	oxygen
H_2O	water
NaOH	sodium hydroxide (caustic soda)

$CBrF_3$	bromotrifluoromethane (BTM or Halon 1301)
HCHO	formaldehyde
$C_2H_5NO_3$	ethyl nitrate

Each compound has its own properties which may be vastly different from those of the constituent atoms. Atoms within a molecule cannot be separated unless the compound undergoes a chemical reaction.

A chemical reaction occurs when the atoms in molecules rearrange either by decomposing into smaller molecules or by joining with other atoms to form different molecules; in both cases the atoms reorganise themselves to form different structures. Endothermic reactions require the input of heat to make them happen whereas exothermic reactions occur with the evolution of heat.

$$2Na + 2H_2O = 2NaOH + H_2 + \text{heat}$$
$$2H_2 + O_2 = 2H_2O + \text{heat}$$

These chemical equations show in chemical shorthand, using the chemical codes, the rearrangement of atoms which occurs in these reactions, a balance of the number of atoms being maintained during the reaction. The molecular mass of molecules can be obtained by adding together the mass numbers of the constituent atoms.

Compounds which contain atoms of elements other than carbon, but including carbon dioxide (CO_2), carbon monoxide (CO) and the carbonates (e.g. calcium carbonate $CaCO_3$) are called inorganic chemicals. All other compounds which contain carbon atoms are known as organic chemicals.

Carbon is an unusual element; not only is it able to form simple compounds where one or two carbon atoms are joined to atoms of other elements, but carbon atoms can link together to form chains or rings of atoms. Almost all other atoms can be joined into these chains and rings to create millions of different organic compounds, from the comparatively simple ones consisting of one carbon with one other type of atom to the highly complex molecules with hundreds of linked carbon atoms joined with other different atoms. Organic chemicals include most of the solvents, plastics, drugs, explosives, pesticides and many other industrial chemical substances.

1.3 Properties of chemicals

Properties of chemicals are to a large extent determined by how the atoms are bonded.

1.3.1 Metals

Metals are different in structure from both types of compounds described below, existing in the solid state as an ordered array of atoms held together by their electrons which circulate freely between them. Applica-

Table 1.1 Properties of typical elements

Element	*Symbol*	*Properties*
Reactive metals		
Aluminium	Al	m.p. 660°C, good conductor, surface oxide formation resists attack by air or water
Barium	Ba	m.p. 850°C, soft, spontaneously flammable in air, reacts with water
Lithium	Li	m.p. 186°C, soft, burns vigorously in air, reacts with water
Less reactive metals		
Cobalt	Co	m.p. 1490°C, hard, not attacked by air or water
Iron	Fe	m.p. 1525°C, burns in oxygen, reacts slowly with water
Mercury	Hg	liquid, slowly attacked by oxygen, no reaction with water
Silver	Ag	m.p. 961°C, ductile, not attacked by oxygen or water
Non-metals		
Bromine	Br	dark red liquid, b.p. 59°C, very reactive, not flammable
Phosphorus	P	red form, m.p. 600°C or white form, m.p. 43°C, burns readily to P_2O_5, insoluble in water
Sulphur	S	yellow or white, m.p. 115°C, burns to SO_2, insoluble in water

tion of an electric potential across a metal allows the electrons to undergo a directional flow between the atoms making metals good electrical conductors. *Table 1.1* lists the properties of some typical metals and other elements.

1.3.2 Inorganic compounds

In some compounds one or more of the bonds joining the atoms are the result of an unequal sharing of electrons between the two atoms and these produce ionic compounds which are crystalline solids, usually with a high melting point. They are often soluble in water giving a solution which conducts electricity.

Many other compounds have bonds based on an equal sharing of electrons and so do not ionise. These compounds can be solids having low melting points, or liquids or gases. Usually they are not soluble in water unless they react with it. There are also many more compounds with types of bond intermediate between the two described and which exhibit properties that relate to both types. *Table 1.2* lists some of the properties of a selection of inorganic compounds.

With the exception of sulphur, stannic chloride and potassium chloride, all the elements and compounds listed in *Tables 1.1* and *1.2* present hazards to health.

Table 1.2 Properties of a selection of inorganic compounds

Compound	*Formula*	*Properties*
Ammonia	NH_3	gas, b.p. −33°C, dissolves readily in water giving a basic solution
Carbon monoxide	CO	gas, b.p. −190°C, odourless, almost insoluble in water
Hydrogen chloride	HCl	gas, b.p. −85°C, dissolves readily in water giving hydrochloric acid
Hydrogen sulphide	H_2S	gas, b.p. −61°C, strong odour, burns in air
Hydrogen peroxide	H_2O_2	liquid, decomposes violently on heating, powerful oxidising agent
Stannic chloride	$SnCl_4$	liquid, b.p. 114°C, fumes, reacts rapidly with water
Sulphuric acid	H_2SO_4	liquid, decomposes at 290°C giving SO_3, strong acid, reacts violently with water
Aluminium silicate	$Al_2Si_2O_7$	solid, infusible, unreactive, clay silicate
Phosphoric acid	H_3PO_4	solid, m.p. 39°C or syrupy liquid, strong acid
Potassium chloride	KCl	solid, m.p. 770°C, very soluble in water, unreactive
Sodium hydroxide	$NaOH$	solid, m.p. 318°C, deliquescent, strong alkali

1.3.3 Organic compounds

As most organic compounds contain a relatively large percentage of carbon and hydrogen atoms they are flammable and many are toxic. All living matter is constructed of complex interdependent organic chemicals and it is because organic compounds interfere with the normal functioning of living matter that they constitute fundamental health and hygiene hazards.

Although there are very many organic compounds they can be grouped into a small number of classes according to their reactive properties. These broad groups are listed in *Table 1.3* which gives examples of compounds in each group.

1.3.4 Acids and bases

Acids are compounds which dissolve in water to give hydrated hydrogen ions:

$$HCl + H_2O = H_3O^+ + Cl^-$$
$$H_2SO_4 + H_2O = H_3O^+ + HSO_4^-$$

Strong acids completely dissociate into ions in solution; weak acids only partially dissociate. A concentrated acid is one which is not diluted with water, and the terms *strong* and *concentrated* should not be confused.

Acids are corrosive in that they react with both metals and with body proteins. Acids are dangerous not just because of their acidity but they can be oxidising agents (HNO_3, $HClO_4$), violently reactive with water

Table 1.3 Examples of the main groups of organic compounds

Group	*Example*	*Formula*	*Use*
Aliphatic hydrocarbons	Methane	CH_4	natural gas
	Butane	C_4H_{10}	petroleum gas
Aromatic hydrocarbons	Benzene	C_6H_6	toxic solvent
	Toluene	$C_6H_5CH_3$	solvent
Halocarbons	Bromomethane	CH_3Br	fumigant
	Trichloroethane	CH_3CCl_3	solvent
Alcohols	Ethanol	C_2H_5OH	'alcohol'
	Glycerol	$C_3H_5(OH)_3$	glycerine
Carbonyl compounds	Formaldehyde (methanal)	HCHO	fumigant
	Benzaldehyde	C_6H_5CHO	manfacturing
	Acetone	CH_3COCH_3	solvent
Ethers	Ethyl ether	$C_2H_5OC_2H_5$	anaesthetic
	Dioxan	$C_4H_8O_2$	solvent
Amines	Methylamine	CH_3NH_2	manufacturing
	Aniline	$C_6H_5NH_2$	manufacturing
Acids	Ethanoic acid	CH_3CO_2H	acetic acid
	Phthalic acid	$C_6H_4(CO_2H)_2$	manufacturing
Esters	Ethyl acetate	$CH_3CO_2C_2H_3$	solvent
Amides	Acetamide	CH_3CONH_2	manufacturing
	Urea	$CO(NH_2)_2$	by-product

(H_2SO_4) and many are toxic. Phenol (C_6H_5OH) is one of the most dangerous acidic organic compounds.

Bases are of two types, solid alkalis such as metal hydroxides which dissolve in water to give hydroxide ions, and gases and liquids such as ammonia and the amines, which liberate hydroxide ions on reaction with water:

$$NaOH + H_2O = Na^+ + OH^- + H_2O$$
$$NH_3 + H_2O = NH_4^+ + OH^-$$

Some of the bases are toxic, many react exothermally with water and all are highly corrosive or caustic towards proteins. Alkalis spilled on the skin penetrate much more rapidly than acids and should be leached out with copious water and not sealed in by attempting neutralisation.

The reaction between an acid and a base is a vigorous, exothermic neutralisation forming a salt. The strength of acids and bases can be measured in terms of hydrogen ion concentration by the use of either meters or test papers, and it is expressed as a pH value on a scale from 0 (acid) to 14 (base). Pure water has a neutral pH of 7.

1.3.5 Air and water

Air and water deserve to be considered separately since they are ever present and are necessary for the operation of many processes and responsible for the degradation of many materials.

Air is a physical mixture of gases containing approximately 78% nitrogen, 21% oxygen and 1% argon. These proportions do not vary greatly anywhere on the earth but there can be additional gases as a result of the local environment: carbon dioxide and pollutants near industrial towns, sulphur fumes near volcanoes, water vapour and salts near the sea etc.

Air can be liquefied and its constituent gases distilled off; liquid nitrogen (b.p. –196°C) has many uses as an inert coolant, liquid oxygen (b.p. –183°C) is used industrially in gas-burning equipment and in hospitals, and argon (b.p. –185.7°C) is used as an inert gas in certain welding processes. Liquid oxygen is highly hazardous as all combustible materials will burn with extreme intensity or even explode in its presence. Combustion is a simple exothermic reaction in which the air provides the oxygen needed for oxidation. If the concentration of oxygen is increased the reaction will accelerate. This effect was experienced in the fire on HMS Glasgow[1].

Water is a compound of hydrogen and oxygen that will not oxidise further and is the most common fire extinguishant. However, caution must be exercised in its use on chemical fires since a number of oxides and metals react energetically with it, in some cases forming hazardous daughter products and in others producing heat and hydrogen which further exacerbate the fire.

1.4 Physical properties

All matter, whether solid, liquid or gas, exhibits properties that follow patterns that have been determined experimentally and are well established and proven. This section looks at some of the factors that influence the state of matter in its various forms.

1.4.1 Temperature

Temperature is a measure of the hotness of matter determined in relation to fixed hotness points of melting ice and boiling water. Two scales are universally accepted, the Celsius (or Centigrade) scale which is based on a scale of 100 divisions and the Fahrenheit scale of 180 divisions between these two hotness points. Because Fahrenheit had recorded temperatures lower than that of melting ice he gave that hotness point a value of 32 degrees. Converting from one scale to the other:

$$(°F - 32) \times 5/9 = °C$$
$$(°C \times 9/5) + 32 = °F$$

Man has long been intrigued by the theory of an absolute minimum temperature. This has never been reached but has been determined as being −273°C. The Kelvin or absolute temperature scale uses this as its zero, O K; thus on the absolute scale ice melts at +273 K.

Devices for measuring temperature include the common mercury in glass thermometer, thermocouples, electrical resistance and optical techniques.

1.4.2 Pressure

Pressure is the measure of force exerted by a fluid (i.e. air, water, oil etc.) on an area and is recorded as newtons per square metre (N/m^2). With solids the term stress is used instead of pressure. Datum pressure is normally taken as that existing at the earth's surface and is shown as zero by pressure gauges which indicate 'gauge pressure' (i.e. the pressure above atmospheric). However, at the earth's surface the weight of the air of the atmosphere exerts a pressure of 1 N/m^2 or 1 bar. Beyond the earth's atmosphere there is no pressure and this is taken as the base for the measurement of pressure in absolute terms. Thus:

$$\begin{aligned} \text{gauge pressure} &= \text{absolute pressure} -1\ N/m^2 \\ \text{or absolute pressure} &= \text{gauge pressure} +1\ N/m^2 \end{aligned}$$

The pressure at the top of a mercury barometer, where the force due to the weight of the atmospheric air outside the tube is balanced by the force exerted by the weight of the column of mercury inside, is normally taken as zero (0 N/m^2 or absolute vacuum), although scientifically there is a small vapour pressure from the mercury.

Pressure can be measured by means of manometers which show the pressure in terms of the different levels of a liquid in a U-tube, by mechanical pressure gauges which record the differential effect of pressure forces on the inside and outside surfaces of a coiled tube or of a diaphragm, and electronic devices which measure the change of electrical characteristic of an element with pressure.

1.4.3 Volume

Volume is the space taken up by the substance. With solids which retain their shape, their volume can be measured with comparative ease. Liquid volume can be measured from the size of the containing vessel and the liquid level. Gases, on the other hand, will fill any space into which they are introduced, so to obtain a measure of their volume they must be restrained within a sealed container.

Each type of material reacts to changes of temperature and, to a lesser extent with solids and liquids, to changes of pressure, by increases or decreases in their volume and this fact can be made use of, or has to be allowed for, in many industrial processes and plant.

1.4.4 Changes of state of matter

At ordinary temperatures, matter exists as solid, liquid or gas but many substances change their state as temperatures change – for example, ice melts to form water at 0°C and then changes into steam at 100°C. The stages at which these changes of state occur are also influenced by the pressure under which they occur.

1.4.4.1 Gases

In gases the binding forces between the individual molecules are small compared with their kinetic energy so they tend to move freely in the space in which they exist. When heated, i.e. additional kinetic energy is given to them, they move much more rapidly and if restrained in a fixed volume impinge more energetically on the walls of the containing vessel, a condition that is measured as an increase in pressure. The relationship between temperature, pressure and volume of gases is defined by the general Gas Law:

$$\frac{PV}{T}\text{(initial)} = \frac{PV}{T}\text{(final)}$$

where P = absolute pressure, V = volume and T = absolute temperature.

Thus in a reaction vessel which has a fixed volume, if the temperature is increased, so the pressure will increase. If the reaction is exothermic and the temperature increase is not controlled there is a risk that the pressure in the vessel could rise above the safe operating level with consequent risk of vessel failure, a situation that may be met in chemical processes that use autoclaves and reactor vessels.

This general law applies with variation when gases are compressed in that the temperature of the gas rises. In air compressors where there is likely to be oil present the temperature of the compressed air must be kept below a certain level to prevent ignition of the contained oil. Conversely, when the pressure of a gas is decreased, the temperature drops, a condition that can be seen with bottles of LPG where a frost rime forms and where in cold weather there is a danger of the temperature of the gas dropping so low that the control valve freezes up.

Some gases can be compressed at normal temperature until they become liquids (e.g. carbon dioxide, chlorine, etc.), and can conveniently be stored in that state, while others, called permanent gases, cannot be liquefied in this way but are stored either as compressed gases (e.g. hydrogen, air etc.) or under pressure in an absorbent substance (e.g. acetylene).

Air, carbon dioxide and a number of other gases which dissolve in water become more soluble as the pressure increases or the temperature decreases. Increase in temperature or decrease in pressure causes the dissolved gases to come out of solution, e.g. tonic water or fizzy lemonade. This is why hydraulic systems need venting. A similar condition arises with divers who surface too quickly and release dissolved gases from their blood causing the 'bends'.

The specific gravity (sp. gr.) of a gas is determined by comparing its weight with that of air and this has important implications at work. The charging of the batteries of fork lift and other trucks results in the generation of hydrogen (which has a sp. gr. of 0.07) which will either become trapped under any covers or lids left over the battery creating a serious explosion risk or will rise into the ceiling space where good ventilation is necessary to prevent a fire risk. At the other end of the scale carbon dioxide (CO_2, sp. gr. = 1.98) and hydrogen sulphide (H_2S, sp. gr. = 1.19) being heavier than air tend to sink to the floor and will fill the lower part of pits and storage vessels making entry hazardous. Similarly, the vapours of most flammable liquids are heavier than air and tend to collect in low places in the floor and will flow like water to the lowest point creating fire hazards possibly remote from the site of the leakage.

1.4.4.2 Liquids

In liquids, the kinetic energy of the molecules is sufficient to allow them to slide over each other but not sufficient for them to move at random in space since they are subject to the binding force of cohesion and remain a coherent mass with a definite volume. Thus the substance will flow to take up the shape of the container but cannot be compressed. When heated the kinetic energy of the molecules increases and the liquid expands until boiling point is reached when the cohesive forces can no longer hold adjacent molecules together, the liquid evaporates and behaves as a gas. If the temperature falls, i.e. the kinetic energy of the vapour is reduced, the vapour will revert to its liquid state by

Table 1.4. Properties of some flammable liquids and gases

Substance	*Sp. gr. of liquid*	*Sp. gr. of vapour*	*Boiling point, °C*	*Sp. heat of liquid at 20°C*	*Vapour pressure at 20°C in mmHg*	*OES long-term ppm*
Acetaldehyde	0.78	1.52	21	–	760	100
Acetone	0.79	2.00	56.5	0.53	185	1000
Acetylene	–	0.91	–84	–	–	*
Ammonia	–	0.60	–33.4	–	–	25
Carbon disulphide	1.3	2.64	46	0.24	298	10 (MEL)
Carbon dioxide	–	1.98	–79	0.82	–	5000
Diethyl ether	0.71	2.56	35	0.54	442	400
Hydrogen	–	0.07	–253	–	–	*
Commercial propane	0.50	1.4–1.55	–45	1.53	9	*
Toluene	0.87	3.14	111	0.39	23	100

OES = occupational exposure standard.
MEL = maximum exposure limit.
* = asphyxiant, requires monitoring for oxygen content of atmosphere.

condensing. However, at temperatures well below the boiling point, some surface molecules gain sufficient energy to escape from the surface. This means that the liquid is constantly losing some of its surface molecules by vaporisation and the extent to which this is occurring is measured as vapour pressure (*Table 1.4*). Vapour pressure will increase as temperature rises.

In passing from the liquid to the vapour or gaseous state a considerable additional input of energy is required without raising the temperature. This energy is the 'latent heat of vaporisation' (H_v in *Table 1.5*) of the liquid and can be quite substantial. This characteristic is made use of in fire fighting where a fine spray of water absorbs the heat of the flames to become steam and in so doing reduces the gas temperature to below that required to maintain combustion.

Table 1.5. Heats of vaporisation and fusion

Substance	*M.P., °C*	*B.P., °C*	*Hv, kJ/kg*	*H_f, kJ/kg*
Aluminium	660	2450	10500	397
Copper	1083	2595	4810	205
Gold	1063	2660	1580	64
Water	0	100	2260	335
Ethanol	−114	78	854	105
Oxygen	219	−183	210	13.8

H_v = heat of vaporisation.
H_f = heat of fusion.

As liquids are heated they expand volumetrically and this effect has to be taken into account in the storage of liquids in closed vessels. It is normal to leave an air space above the liquid, the ullage, to allow for this expansion, and to fit pressure relief valves to release any excess pressure generated. An example of what can happen if these precautions are not taken was seen at Los Alfaques Camping Site at San Carlos de la Rapita in Spain in July 1978 when a road tanker carrying 23.5 tonnes of propylene exploded killing 215 people. The circumstances of the incident were that the unlagged tanker was filled completely with propylene during the cool of the early morning, leaving no ullage. The tank was not fitted with any form of pressure relief. As the tanker was driven in the heat of the sun, the tank and its contents heated up to a temperature where the vapour pressure of the contents exceeded atmospheric pressure and the propylene (an incompressible liquid), expanding at a greater rate than the tank, caused the tank to rupture as it was passing a holiday camp at 14.30 hours creating a BLEVE (boiling liquid expanding vapour explosion) that caused devastation over an area of five hectares.

The incompressibility of liquids is utilised in many ways in industry particularly in hydraulic power transmission. Two common applications are as power sources for machines and vehicles and in hydraulic cylinders. Normally the hydraulic medium is an oil and hazards can arise where leaks occur either as oil pools on a floor or from high-pressure lines as a fine mist which is highly flammable.

1.4.4.3 Solids

In a solid at room temperature the molecules are tightly packed together, have little kinetic energy and vibrate relative to each other. A solid has form and a rigid surface as a result of the strength of the cohesive forces between the molecules. As a solid is heated the vibrations of the molecules increase and the solid expands. Eventually a temperature is reached where the kinetic energy allows the molecules to slide over one another and the solid becomes a liquid.

Solids, and particularly metals, have a number of characteristics that are of great value in almost everything to do with modern standards of living. They have stability of form over a wide range of temperatures which enables them to be formed and machined into shapes, a measurable strength over a wide range of stresses and temperatures, and permanent physical characteristics such as rates of expansion and contraction with temperature.

The change of state from solid to liquid requires the input of a considerable quantity of heat – the latent heat of fusion (H_f in *Table 1.5*) – without raising its temperature. This can easily be seen with melting ice where the temperature of the water remains at 0°C until all the ice has melted. At room temperatures some solids, such as solid carbon dioxide, do not pass through the liquid stage but change directly from a solid to a gas, a process known as 'sublimation', the liquid state only occurring at low temperature and/or high pressure. Organic substances such as wood also have no liquid stage but decompose on heating, rather than melting, reducing their very large molecules into smaller ones some of which are given off as vapours.

As solids are heated they expand at rates determined by their 'coefficients of expansion' that vary with the different substances. The heat needed to raise the temperature of different solids varies and can be determined from its 'specific heat'. Problems can arise where dissimilar metals are welded together that have different specific heats and different coefficients of expansion. Allowance has to be made for differential expansions where different metals are moving in contact and have different coefficients of expansion, such as in bearings. However, use is made of these different expansion rates in bimetal strips for temperature measurements and in thermostats.

Physical properties of some solids are shown in *Table 1.6*.

Solids, and particularly metals, have great strength and an ability to withstand high levels of stress and strain. By alloying metals they can be made to exhibit particular characteristics to suit particular applications. Characteristics of metals can vary from very ductile lead and gold through high-strength high-tensile steels to brittle cast iron, each having its particular use. One exception is the metal mercury which is a liquid at room temperature.

Table 1.6. Physical properties of some solids

Material	*Density kg/m³*	*Coefficient of linear expansion µm/mK*	*Specific heat kJ/kgK*
Aluminium	2700	24	900
Brass	8500	21	380
Cast iron	7300	12	460
Concrete (dry)	2000	10	0.92
Copper	8930	16	390
Glass	2500	3–8	0.84
Polythene	930	180	2.20
Steel	7800	12	480

1.5 Energy and work

There are two types of energy, kinetic and potential. Kinetic energy is involved in movement and may be available to do work; it can be present in a number of forms such as heat, light, sound, electricity and mechanical movement. Potential energy is stored energy and usually requires an agent to release it, as, for instance, the potential energy of a loose brick lying on a scaffold which needs to be kicked off to produce kinetic energy.

Heat is a form of energy where the degree of hotness of a material is related to the rate of movement of the atoms and molecules which make it up. Heat is transferred from one material to another by conduction, convection or radiation. Conduction occurs within and between materials that are in contact through the physical agitation of molecules by more energetically vibrating neighbours. Some materials are better heat conductors than others depending on the ease with which molecules can be made to vibrate. Convection is the conveyance of heat in gases and liquids by the heated fluid rising and heating any surface with which it comes into contact. Initial spread of fire within a building is most likely to be by convection and a method of control is by venting the hot gases to outside the building so preventing the lateral spread of fire. Radiation of heat from a heat source occurs as infrared emissions which pass through the atmosphere by wave propagation.

If a force is exerted on an object, no energy is expended until the body moves when the force is converted into kinetic energy which is equal to the work done on the body by the force. Mechanical work is done whenever a body moves when a force is applied to it.

$$\text{Work done (joules)} = \text{force (newtons)} \times \text{distance moved (metres)}$$

In all mechanical devices or machines the work done is never as great as the energy expended, as some of the energy is lost in overcoming friction. The ratio of work done to energy expended is the mechanical efficiency of the machine and is always less than one.

Power is the rate of doing work and is measured in joules per second:

$$1 \text{ joule per second} = 1 \text{ watt of power}$$
$$1\,\text{kWh} = 3.6 \times 10^6 \text{ joules}$$
$$746 \text{ watts} = 1 \text{ horsepower}$$

The rate at which machines work is given in either horsepower or kilowatts.

Pressure is unreleased potential energy since when contained in a pressure vessel the fluid exerts equal and opposite forces on the vessel walls but no movement takes place. Pressure is independent of the shape of the vessel and is exerted at right angles to the containing surfaces. With gases the pressure is virtually the same throughout the containing vessel but with a liquid the pressure varies according to the depth, i.e. the weight of liquid above the point of measurement. Pressure can be converted into kinetic energy through the movement of a piston in a pneumatic or hydraulic cylinder.

1.6 Mechanics

Mechanics is that part of science that deals with the action of forces on bodies. Much of the theory is based on Newton's three laws of Motion:

1 Every body continues in its state of rest or of uniform motion in a straight line except in so far as it is compelled by external impressed force to change that state.
2 Rate of change of momentum is proportional to the force applied and takes place in the direction in which the force acts.
3 To every action there is always an equal and contrary reaction.

If a force is applied to a body resting on a plane, initially the body will not move because of the friction between itself and the plane. The force is resisted by an equal and opposite force due to friction. Once the applied force exceeds the friction force the body will move in the direction of the applied force. However, if the same force continues to act on the body its speed will accelerate because sliding friction is less than limiting (i.e. static) friction. Hence:

Before movement the force $F = W\mu_L$

where μ_L is the coefficient of static or limiting friction, and W is the reaction between the body and the plane (i.e. its weight = mass × gravity).

After movement the excess force $F_e = F - W\mu_s$

where μ_s is the coefficient of sliding friction.

This excess force overcomes the *inertia* of the body and causes it to accelerate at a rate f given in the formula

$$F_e = mf$$

This is why a sticking object will shoot away once it has been freed. Friction has an ambivalent role – it is necessary to enable us to walk, for cars to move and to enable us to control motion through the use of brakes; on the other hand in machines friction absorbs energy and makes the machine less efficient.

Bodies can possess potential energy as a result of the height at which they are located above a datum and that potential energy can be converted into kinetic energy by the object falling. Thus a mass m falling through a height h does work equal to

$$\begin{aligned} W &= m \times g \times h \\ &= \tfrac{1}{2} \times m\, v^2 \text{ its kinetic energy} \end{aligned}$$

where v is its velocity.

For a body to move in the direction of the force applied to it, the line of that force must pass through the centre of gravity of the body. If the line of the force does not pass through the body's centre of gravity, the force will apply a turning moment to the body. Two equal and opposite forces acting on a body and not passing through its centre of gravity will apply a turning couple.

In machines, forces act on component parts in overcoming friction, accelerating or decelerating or in transmitting power or loads. These forces can be tensile, compressive, shear, bending or torsional.

1.7 Strength of materials

For machines, plant and buildings to work and give an economic life, their component parts must be capable of resisting the various forces to which they are subject during normal working. Any load applied to a component induces stresses and strains in it. The type of stress depends on the manner in which the load is applied. Stress is force per unit area (N/m^2). Strain is the proportional distortion when subject to stress ($\delta l/l$) and is often quoted as a percentage. The ratio of stress to strain is constant and known as Hooke's Laws after its discoverer, thus:

$$\frac{\text{stress}}{\text{strain}} = \text{constant} = E$$

E is called Young's modulus or modulus of elasticity and its value depends on the material and the type of stress to which it is subjected.

Forces acting on a component can be tensile, compressive, shearing, bending or torsional. The effect of these is shown diagrammatically in *Figure 1.2*.

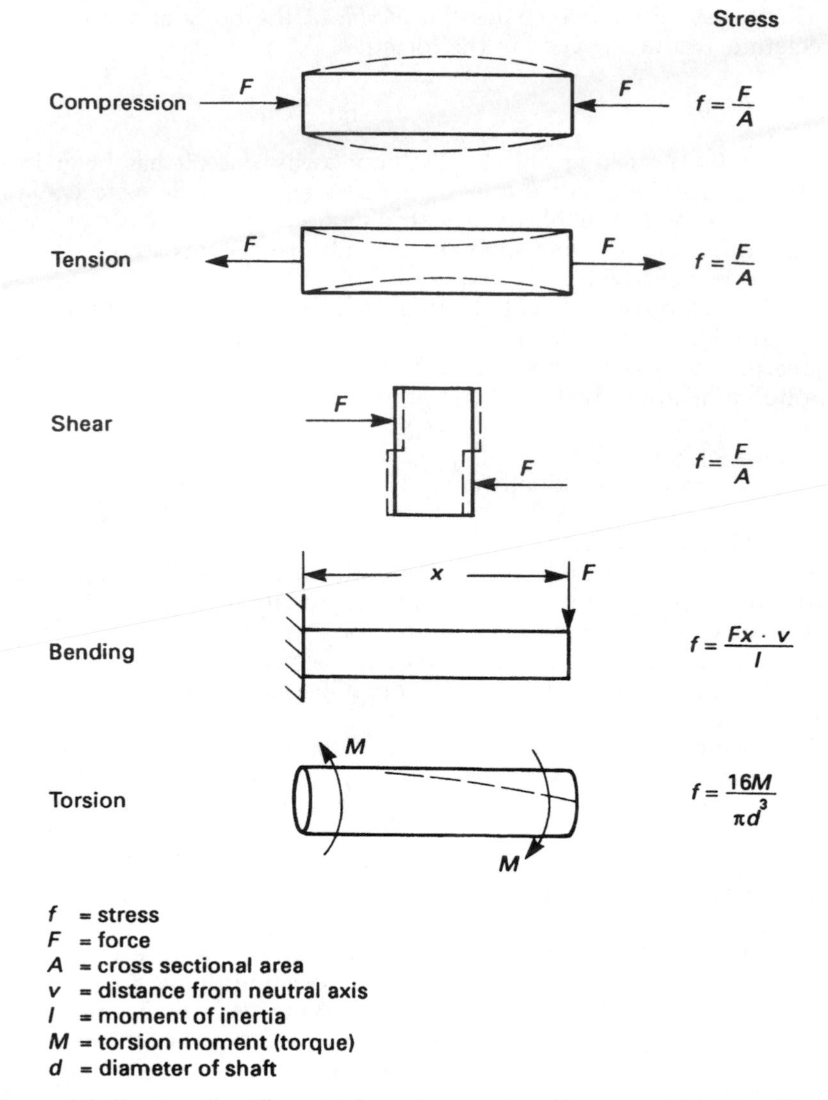

Figure 1.2 Showing the different deformations produced in a material by the different forces acting on it

The calculation of tensile, compressive and shear stresses is relatively easy, but to obtain bending stresses it is necessary to know the moment of inertia of the section of the beam and in the case of torsion, although it is a shear stress, because it is spread along the length of the shaft its calculation is complex. With bending moments, a tensile stress is induced in the outer surface of the beam and an equal and opposite compressive stress occurs in the inner surface.

While each material exhibits particular physical characteristics, they all follow a broad pattern of behaviour as shown in *Figure 1.3*. Where an item

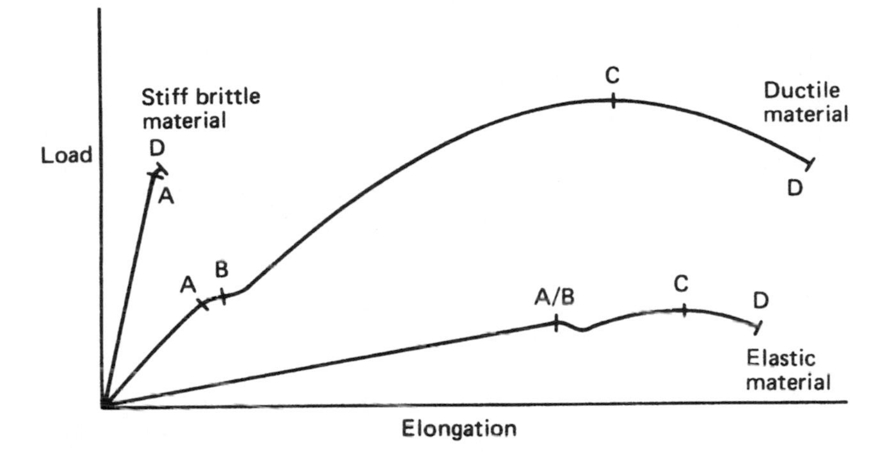

Figure 1.3 Behaviour of different materials under tensile load. A = elastic limit; B = yield point; C = ultimate tensile strength; D = breaking point

has a load applied to it, provided that the strain induced does not go beyond the elastic limit, then when the load is removed the item will return to its original size, i.e. there will have been no permanent deformation However, if the elastic limit is exceeded permanent deformation occurs and the characteristic of the material will have changed. Typical stiff brittle materials are cast iron, glass and ceramics which have virtually no ductility and will fracture at the elastic limit. Ductile materials include mild steel and copper, and elastic materials include plastics.

1.8 Modes of failure

Even in the best of designed machinery and plant failures can occur. When they do, the cause is normally investigated to determine the mode of failure so that repetition can be prevented. Modes of failure relate to the type of stress to which the component has been subjected and the characteristic features of failures due to tension, compression, shear and torsion are well known. Sometimes the failure is related more to the operating process than to the stress, particularly when there is repeated stress cycling of the part when it can suffer fatigue failure.

In processes using certain chemicals, stresses have been shown to make the parts prone to corrosive attack which can reduce their strength. Similarly failures have occurred where a chemical in contact with the part has affected its ability to carry stresses through causing embrittlement such as zinc embrittlement of stainless steel[2] and hydrogen embrittlement of grade T chain[3].

1.9 Testing

While there are common legislative requirements for the testing of finished products where, in normal work, the equipment is stressed, such as pressure vessels, cranes, lifting equipment etc., a great deal of testing takes place long before the product is manufactured. It starts at the material producer who tests his product for quality and to ensure that it meets specification. Such materials testing is usually on specially shaped test pieces that are tested to destruction and by chemical analysis to ensure that the constituent materials are present in the correct amounts. The supplier will issue the appropriate test certificates.

In the manufacture of plant and machinery, and following subsequent periodic inspections, should they become necessary testing is normally non-destructive and a number of specialised techniques have been developed. The aim is to check the condition of the material of which the plant is constructed and to identify faults that cannot be seen by eye. Special detection techniques used to highlight the weaknesses or faults in the material include the use of magnetic particles, penetrant dyes, X-ray and gamma-ray sources, ultrasonic vibrations, microwave and infrared rays. Equipment is now available to enable visual inspections to be carried out of inaccessible places using fibre optics and remote-controlled television.

1.10 Hydraulics

This side of engineering is concerned with fluids, both liquids such as water, oil etc., and gases such as air and other gaseous elements and compounds. Steam can be classified as a gas although it has characteristics that make it more hazardous than normal compressed gases. These two types of fluid exhibit disparate characteristics in that gases can be compressed while liquids are incompressible.

1.10.1 Compressible fluids

When gases are compressed, work is done on them and they acquire considerable potential energy as pressure. They also become hot and it is sometimes necessary to provide cooling to ensure that critical temperatures are not exceeded. Conversely when gas pressure is released, the gas does work or gives up energy and heat and becomes cold. This latter effect is the basis by which refrigerators work and it can be seen, even in the middle of summer, as rime on the outside of LPG cylinders.

Compression of gases and air to achieve high pressures with relatively low volumes is by positive displacement compressors of either the piston or rotary type, while for large volumes at lower pressures centrifugal compressors are used as for example in gas turbines. Gas flows can be measured by means of a vane or hot wire (heated head) anemometer, by measuring the pressure difference across an orifice[4], or through the use of pitot tubes.

Compressed gases have a large amount of potential energy which can be utilised for actuating thrust cylinders. Once pressurised gas has been admitted to a cylinder, even if the supply is then cut off, the cylinder will continue to operate until all the energy of the gas has been absorbed in work done. A dramatic demonstration of the energy of compressed gases can be seen when a containing vessel ruptures with heavy sections being projected very considerable distances or the jet effect when a valve breaks off a gas cylinder. Because of the energy it contains, compressed air is not used for pressure testing of vessels except in certain very specific cases where the use of water is not practicable, e.g. in nuclear reactor vessels.

1.10.2 Incompressible fluids

Normally referred to as liquids, incompressible fluids have many advantages when used for power transmission in that, even at very high pressures, they do not in themselves contain a great deal of energy so that should a failure of a containing vessel or pipework occur the result is not catastrophic. These fluids also have the advantage that their flow and pressure can be controlled to a very fine degree which makes them ideal for use in applications where control is critical. They also have the advantage that they can be pumped at very high pressures and so permit the application of very large forces through hydraulic jacks and high torques through hydraulic motors. The power of a fluid jet has been harnessed in cutting lances where very high-pressure liquid, normally water, is emitted from a very small orifice with sufficient force to cut through a variety of materials including steel. These lances have a number of applications but can be very dangerous so their use requires strict control.

The pumps used in handling these high-pressure liquids can suffer considerable damage from cavitation. Incompressible liquids will not compress, nor will they withstand tension; thus if the suction inlet to a pump is restricted the fluid will release any contained air to form cavities. This condition seriously affects the performance of the pump, can cause damage to its rotor and generates a great deal of noise. Gas or air entrained in a hydraulic fluid is detrimental to its effectiveness as a power transmission medium, an effect that may be experienced in the braking system of a vehicle.

1.11 Summary

Although much of occupational safety is now recognised as being behavioural, engineering with its foundation in science still has a major role to play. Many of the scientific principles learnt in the lecture room and scientific laboratory have application across a spectrum of work activities and have important connotations as far as occupational safety is concerned. This chapter has reviewed some of those applications – there are many more for the student to discover.

References

1. Health and Safety Executive, *Investigation report: Fire on HMS Glasgow, 23 September 1976*, HSE Books, Sudbury (1978)
2. Health and Safety Executive, *Guidance Note, Plant and Machinery series No. PM 13: Zinc embrittlement of austenitic stainless steel*, HSE Books, Sudbury (1977)
3. Health and Safety Executive, *Guidance Note, Plant and Machinery series No. PM 39: Hydrogen embrittlement of grade T chain*, HSE Books, Sudbury (1984)
4. British Standards Institution, BS 1042: *Measurement of fluid flows in closed conduits, Part 1: Pressure differential devices and Part 2: Orifice plates, nozzles and venturi tubes inserted in circular cross section conduits running full*, BSI, London.

Reading list

Basic Engineering Science

Titherington, D. and Rimmer, J.G., *Engineering Science*, Vol. 1 (1980), and Vol. 2 (1982), McGraw-Hill, Maidenhead

Deeson, E., *Technician Physics*, Longman, Harlow (1984)

Harrison, H.R. and Nettleton, T., *Principles of Engineering Mechanics*, 2nd edn, Edward Arnold, London (1993)

Hannah, J. and Hillier, M.J., *Applied Mechanics*, 3rd edn, Pitman, London (1995)

Houpt, G.L., *Science for Mechanical Engineering Technicians*, McGraw-Hill, Maidenhead (1970)

Properties of Materials

Peapell, P.N. and Belk, J.A., *Basic Materials Studies*, Butterworth, London (1985)

Timings, R.L., *Engineering Materials*, Vol. 1, Longman, Harlow (1998)

Bolton, W., *Engineering Materials Technology*, 3rd edn, Butterworth-Heinemann, Oxford (1998)

Crane, F.A.A. and Charles, J.A. *Selection and Use of Engineering Materials*, 2nd edn, Butterworth, London (1989)

Hanley, D.P., *Introduction to the Selection of Engineering Materials*, van Nostrand, Wokingham (1980)

Young, W.C., *Roark's Formulas for Stress and Strain*, 6th edn, McGraw-Hill, New York (1989)

Higgins, R.A., *Properties of Engineering Materials*, 2nd edn, Hodder and Stoughton, Sevenoaks (1994)

Gordon, J.E., *The New Science of Strong Materials*, 2nd edn, Pitman, London (1991)

Hydraulics

Turnbull, D E., *Fluid Power Engineering*, Butterworth, London (1976)

Pinches, M.J. and Ashby, J.G., *Power Hydraulics*, Prentice-Hall, Hemel Hempstead (1996)

Massey, B.S., *Mechanics of Fluids*, 6th edn, van Nostrand Reinhold, London (1989)

Basic Introductory Chemistry

Lewis, M. and Waller, G., *Thinking Chemistry*, Oxford University Press, Oxford (1986)

Jones, M., Johnson, D., Netterville, J. and Wood, J., *Chemistry, Man and Society*, Holt, Rinehart and Winston, Eastbourne (1980)

Hazardous Chemicals

Schieler, L. and Pauze, D., *Hazardous Materials*, van Nostrand Reinhold, Wokingham (1976)

Hazardous Chemicals, A manual for schools and colleges, Oliver and Boyd, Edinburgh (1979)

Chapter 2

Fire precautions

P. Waterhouse and revised by Ray Chalklen

2.1 Introduction

The control of fire provided man with his first means of advancement. It broadened his food choice by enabling him to cook, it widened his living range by providing him with an external source of heat, it improved his tools by permitting him to extract and work with metals and it lengthened his day by giving him a source of light. Fire was so important to primitive man that he made it one of the four 'elements' – earth, fire, air and water – which made up his world. However, man's discovery of fire also revealed to him its awesome destructive power. Primitive man used fire and worshipped it but he also went in fear and dread of its destructive nature.

Fire precautions are the measures taken in the provision of the fire protection in a building or in other situations to minimise the risk to the occupants, contents and structure from an outbreak of fire.

Fire prevention is the concept of preventing outbreaks of fire, of reducing the risk of fire spreading and of avoiding danger to persons and property from fire.

Fire protection deals with design features, systems or equipment in a building, structure or other fire risk situation, to reduce the danger to persons and property by detecting, extinguishing or containing fires.

2.2 Basic combustion chemistry

When a fire occurs, products of combustion, often called smoke, are evolved and heat and light are given off. Scientifically the process is known as combustion and the reaction that occurs is a gaseous exothermic oxidation reaction. Exothermic means that energy is given off when the reaction occurs, oxidation means that it is a reaction between the substance and oxygen and gaseous that it takes place in the gaseous state. In chemical terms the reaction is written as the following equation:

substance + oxygen → oxidation products – energy

e.g. $CH_4 + 2O_2 = 2H_2O - 890\,kj$

Note: The evolution of energy is always expressed as a negative number. The amount is a characteristic of the reacting substance (the fuel).

Useful as it is to express combustion as a chemical reaction, the equation contains no information on the energy that is required to start it, how fast it will proceed once it has started and the conditions that are necessary to sustain it once it has started. For a reaction to take place it is necessary to put some energy into the mixture of reactants, the initiation energy, usually in the form of heat. Some mixtures require very little heat but for others a large amount is necessary. Often the heat input comes from heating the mixture but the heat can also arise from an electric discharge taking place in the mixture. In fire terms the initiation energy is known as the ignition energy and the source is called the ignition source.

The speed of the reaction is influenced by many factors such as the chemical nature of the combustible substance, the amount of mixing and the physical state of the substances but for an exothermal reaction increasing the temperature of the reactants will increase the speed of the reaction. The difference between a fire and an explosion is very largely the speed of the reaction that takes place once it has been initiated.

The conditions to sustain a combustion reaction once it has started are more difficult to explain in simple terms. At its simplest it is useful to imagine a chemical reaction occurring when two different molecules, A and B, collide, fuse together and then separate as two different molecules, C and D, with the release of some energy. If there are plenty of both A and B molecules around then collisions will occur frequently and the reaction will be seen as taking place. If, however, there are a large number of A molecules and very few of B then B will be quickly used up and the reaction will cease. Similarly if there are few A molecules but a large number of B, the A molecules will be quickly used up and the reaction will again cease. The measure of the number of molecules in a mixture is the concentration of the substance, the higher the number of molecules the greater the concentration. It follows therefore that it is not the fact that oxygen and a combustible material are present but rather the concentration in air (which contains the oxygen) of the combustible substance that is important. In particular there are two concentrations, an upper and a lower, beyond which no reaction will take place. In combustion terms these are known as the upper and lower flammable limits.

2.2.1 The fire triangle

The simple facts of combustion chemistry, reactants and ignition source, are commonly displayed as the fire triangle (*Figure 2.1*).

It is important to note that the fire triangle contains no information on the speed of the reaction or of the amount of energy given off or of the upper and lower flammable limits.

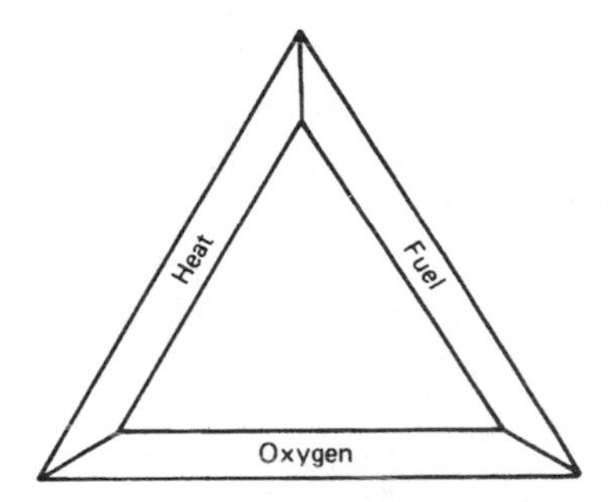

Figure 2.1 Fire triangle. The sides of the triangle represent the elements of combustion

2.2.2 Fuel

Almost all organic chemicals and mixtures whether in the solid, liquid or gaseous state are flammable so they are potential fuels. Examples of these include: wood, paper, plastics, fibres, petrol, oil, LPG, coal, and living tissue. Most but not all inorganic substances are not flammable and hence are not potential fuels. Exceptions to this are: hydrogen, sulphur, phosphorus, magnesium, titanium and aluminium.

2.2.3 Oxygen

The main source of oxygen in a fire is air which contains 21% oxygen. Other sources are oxidising agents and combustible substances which contain oxygen. Examples of the former include organic peroxides, hydrogen peroxide, sodium chlorate and nitric acid, and of the latter ammonium nitrate.

2.2.4 Means of ignition

The most common means of ignition is heat from an external source such as a burning match, a fire, a hot bearing, friction, incandescent sparks or the heat from a light bulb. Electrical discharges from the making and/or breaking of an electric circuit or from static electricity are also possible means of ignition.

2.2.5 Combustion characteristics

For liquids and gases it is possible to measure some combustion characteristics, i.e. the ignition energy and the flammable limits. When the measurements are carried out under standard conditions comparisons can be made and conclusions drawn.

There are three measures of ignition energy, the flash point, the fire point and the ignition temperature. These are defined as:

2.2.5.1 Flash point

The flash point is the minimum temperature at which sufficient vapour is given off to form a mixture with air which is capable of ignition under prescribed test conditions.

2.2.5.2 Fire point

The fire point is the lowest temperature at which the heat from the combustion of a burning vapour is capable of producing sufficient vapour to sustain combustion.

2.2.5.3 Ignition temperature

The ignition temperature is the lowest temperature at which the substance will ignite spontaneously.

The fuel/oxygen criteria are known as the lower and upper flammable (explosive) limits.

In all cases the flash point has a lower value than the fire point and the fire point has a lower value than the ignition temperature. The reason for this is that with the fire point the energy for ignition is provided from an external source and with the ignition temperature it is provided entirely by the heat contained in, i.e. the temperature of, the combustible substance.

The flammable limits are defined as:

2.2.5.4 Lower flammable (explosive) limit

The lower flammable (explosive) limit is the smallest concentration of flammable gas or vapour in air which is capable of ignition and subsequent flame propagation under prescribed test conditions.

2.2.5.5 Upper flammable (explosive) limit

The upper flammable (explosive) limit is the greatest concentration of flammable gas or vapour in air which is capable of ignition and subsequent flame propagation under prescribed test conditions.

It should be noted that, by virtue of the definition, flash point and fire point apply only to liquids while ignition temperature and flammable limits apply to both liquids and gases.

The standard tests for determining the flash point of a liquid are described in Schedules 1 and 2 of the Highly Flammable Liquids and Liquefied Petroleum Gases Regulations 1972 and are referred to in part III of Schedule 1 of the Chemicals (Hazard Information and Packaging for Supply) Regulations 1994 (CHIP 94). The properties of some flammable substances are given in *Table 2.1*.

Table 2.1. Properties of some flammable substances

Substance	*Flash point (°C)*	*Ignition temperature (°C)*	*Flammable range (% v/v in air)*
Acetic acid	40	485	4–17
Acetone	−18	535	2.1–13
Acetylene	−18	305	1.5–80
Ammonia	Gas	630	1.5–27
Benzene	17	560	1.2–8
n-Butane	−60	365	1.5–8.5
Carbon disulphide	−30	100	1–60
Carbon monoxide	Gas	605	12.5–74.2
Cyclohexane	−20	259	1.2–8.3
Ether	−45	170	1.9–48
Ethanol	12	425	3.3–19
Ethylene	Gas	425	2.7–34
Hydrogen	Gas	560	4.1–74
Methane	Gas	538	5–15
Toluene	4	508	1.2–7
Vinyl chloride	−78	472	3.6–33

2.3 The combustion process

Although easy to describe in chemical terms, the process of combustion is complex particularly for liquids and solids. Consider first a pool of liquid. Immediately above the liquid surface there is a layer of vapour, the concentration of which depends on the vapour pressure of the liquid at the temperature in question. If this is ignited and it continues to burn, the heat from the flame causes more of the liquid to evaporate and the burning is sustained. Close examination of the surface of the liquid shows that there is a thin gap between the liquid surface and the flame. In this gap the vapour concentration is above the upper flammable limit. With a solid there is no vapour above the solid to ignite so initially the heat from a flame raises the surface temperature of the solid which causes some decomposition of the surface layer with the evolution of a flammable vapour. This ignites, the heat from the flame causes more decomposition and the burning is sustained. Again, as with the liquid, close examination of the surface of the solid shows that there is a thin gap between the surface and the flame. In this gap the vapour concentration is above the upper flammable limit.

The chemical equation always shows that energy is released and that colourless gases are formed. The energy is actually the heat of the products of combustion and infrared radiation. In reality however, when combustion occurs there is a visible flame and there is smoke. The visibility of the flame arises from glowing particles, usually particles of carbon, which have not been fully oxidised to carbon dioxide. In addition to this visible radiation there are also infrared and ultraviolet radiations.

Smoke is particulate matter where the particles are not sufficiently hot to glow and hence appear black. Smoke particles tend to be larger than the glowing particles in a flame and may include other decomposition substances. The visibility of the combustion flame differs with different substances, for example methanol burns with an almost invisible flame whereas toluene produces a dense black smoke. Generally the more air there is in a fire the less dense will be the smoke.

The hot products of combustion rise and will increase the temperature of everything with which they come into contact. Radiation is emitted uniformly in all directions – infrared radiation will heat any surface on which it falls. The intensity of the radiation diminishes as the square of the distance from the point of emission.

2.3.1 Flammable liquids

These liquids can be divided into two groups, flammable and highly flammable. Liquids with a flash point between 32°C and 55°C are classed as flammable and those with a flash point below 32°C are classed as highly flammable. When a liquid is emitted as a spray, either deliberately through some form of atomiser or inadvertently leaks out under pressure through a small hole, its exposed surface area is increased which allows it to heat up more quickly. Also the unit mass of each droplet is reduced which, again, allows it to heat up more quickly thus increasing the evaporation rate. With the increase in evaporation there is more fuel to burn in the vapour phase and so burning is enhanced. In fact, burning liquid sprays exhibit many of the characteristics of burning gas.

2.3.2 Flammable solids and dusts

Solids do not have the divisions of flammability that liquids do. Dusts, however, are finely divided solids and, as with sprays, they have an increased surface area and a decreased unit mass. Both of these features allow the particles to heat up more rapidly and so their rate of burning is increased. Dusts exhibit many of the burning characteristics of gases.

It is possible to measure the flammability limits of both dusts and sprays and to compare combustion characteristics under standard conditions (see *Table 2.2*).

2.3.3 Spontaneous combustion

A combustion characteristic which appears to be outside the usual range of characteristics is spontaneous combustion. This is the phenomenon of the apparently inexplicable bursting into flame or smoke of some substances such as oil-soaked rags or overalls, oil-soaked lagging, hay and straw. The explanation is simple: a slow oxidation of the oil gives off heat with the formation of substances with lower molecular weights which also have lower flash points. As the oxidation continues the

Table 2.2. Combustion characteristics of various dusts of particle size not more than 200 mesh

Substance	*Ignition temperature (°C)*	*Minimum spark ignition energy (mJ)*	*Lower explosive concentration (mg/m³)*
Magnesium	520	40	200
Zinc	600	650	4800
Methyl methacrylate	440	15	200
Polystyrene	490	15	150
Coffee	160	85	500
Sugar	350	30	350
Coal	610	60	550
Wood floor	430	20	400

temperature of the oxidising oil reaches the flash point of one of the degradation products and a fire occurs. The oxidation and generation of heat accelerate if the heat of oxidation cannot escape. The reaction rate is also increased when the oil is spread over a large surface. With hay and straw the initial source of heat is bacterial decomposition initiated by wet conditions and as hay and straw are good heat insulators the conditions for acceleration of the reaction are always present.

2.4 Classification of fires

Fires are commonly classified into four categories which take into account the type of substance that forms the fuel and the means of extinction.

2.4.1 Class A

Fires involving solid materials normally of an organic nature such as wood, paper, natural fibres etc. in which combustion occurs with the formation of glowing embers. Water is the most effective extinguishing agent and the fire is extinguished by cooling the burning material.

2.4.2 Class B

Fires involving liquids or liquefiable solids such as petrol, oil, fats etc. The most effective extinguishing agents are foam and dry powder which extinguish the fire by excluding the oxygen.

2.4.3 Class C

Fires involving gases or liquefied gases such as methane, propane, butane etc. Fires of this type can be difficult to extinguish, and if they involve gas should not be extinguished until the gas supply has been shut off as an explosive mixture can quickly build up from escaping gas. Dry powder is the best extinguishing medium.

2.4.4 Class D

Fires involving metals such as magnesium, aluminium, sodium and potassium can only be extinguished by special dry powders which include powdered graphite, powdered talc, soda ash and dry sand and all must be applied carefully to ensure they cover the burning material. These extinguishers work by smothering the fire.

2.4.5 Electrical fires

There is no separate classification for electrical fires since any of the previous classes of fire could be started by faulty electrical equipment. Carbon dioxide is the most effective extinguishing medium for fires involving electrical equipment but the electricity supply must be turned off before the fire is tackled.

2.4.6 Extinguishing agents

Continued development of fire extinguishing agents and methods of application have led to the evolution of multipurpose extinguishers. Manufacturers' claims for firefighting capabilities should be carefully checked against the likely risks for the area where the extinguisher is to be used.

2.5 Ignition sources and their control

2.5.1 Sources of ignition

As heat is the most common source of ignition anything which will supply heat to a flammable substance is a potential ignition source. Direct heating sources such as steam and hot-water pipes, hot air, stoves and boilers, open fires, electrical heaters are just some of a long list of heat sources. Other common sources are friction heating and the hot surfaces of machines. Another source of ignition is the electrical discharge which can take place when a circuit is made or broken, also when static electricity is discharged to earth. Smoking is also a common cause of fire and should only be allowed in areas that are clearly marked and provided with suitable extinguishing receptacles. When hot work, such as grinding, welding or flame cutting, has been taking place, the work area should not be left unattended for at least 30 minutes after work has ceased and whenever possible should be monitored at regular intervals over the next few hours.

2.5.1.1 Heating systems

Hot pipes and ducts should be lagged where they pass close to combustible materials. Heaters of all types should be securely fixed to walls or floors and should be provided with a suitable fence or guardrail to prevent combustible materials from being placed too close to them.

2.5.1.2 Friction

Heating can occur when two materials, such as the parts of a machine, are in direct rubbing contact. The heat generated by friction can be transmitted by both conduction and convection to combustible material nearby. Regular maintenance and correct lubrication will reduce the likelihood of overheating due to friction.

2.5.1.3 Hot surfaces

Hot spots occur on many machines used in industry and commerce but these occurrences can be reduced to a minimum by good standards of maintenance. Where heat is an integral part of the process, either suitable insulation should be provided or other measures taken to ensure that flammable materials are kept away from the areas of high temperature.

2.5.1.4 Electricity

Electricity as such is not a source of ignition, it is the arcing which occurs when a circuit is made or broken, or the heat generated by the passage of electricity through a conductor, that creates the ignition source. The energy of an arcing discharge can be reduced by special measures and this approach is particularly important in potentially flammable atmospheres. An incandescent bulb can create sufficient heat to ignite flammable material placed too close to it and precautions should be taken to ensure that stored goods are not in close contact with such bulbs (*Figure 2.2*).

2.5.1.5 Static electricity

Static electricity is generated where there is relative motion between, and/or separation of, two dissimilar materials. The best method of control is to prevent the charge from building up in the first instance, by providing an earthing link or through the use of static eliminators.

2.5.1.6 Internal combustion engines

All internal combustion engines have hot surfaces and generate hot exhaust gases. In addition petrol engines require sparks to make them

Figure 2.2 Tungsten bulb too close to packing case creating fire risk. (Courtesy Fire Protection Association)

work and most engines have electrical equipment on them. In places where there is a possibility of the presence of flammable gases or vapours the use of petrol-driven engines should be banned. It should not be forgotten that flammable vapours may be sucked into internal combustion engines along with the air they require and this can cause the engine to run out of control.

2.5.1.7 Tools

In areas where flammable vapours or combustible materials may be present, the use of tools that generate hot surfaces, use heat for their operation or give off sparks should be strictly controlled. Where flame cutting, welding or grinding are carried out, the working area should be clearly defined and, where possible, surrounded by a non-flammable curtain. A standby fireman should be present during the operation and for at least 30 minutes afterwards.

2.5.1.8 Smoking

Statistics indicate that smoking is a major cause of fire. The main problem lies with the disposal of the sources of ignition – the match and the cigarette end. Banning smoking can drive it underground and the problem is better controlled by providing designated smoking areas where proper ash trays and waste bins should be available (*Figure 2.3*).

Figure 2.3 Smoking bay in high fire risk factory

2.5.2 Control of flammable substances

Since all organic and some inorganic substances are flammable it is virtually impossible to eliminate them from the workplace. However, they can be controlled.

2.5.2.1 Waste, debris and spillage

These are a frequent source of fire both in the workplace and also when stored outside a building especially if in a combustible container. The risk of fire can be controlled by maintaining high standards of housekeeping. Flammable waste, whether solid or liquid-soaked rags, should be put in a non-flammable container with a covering lid which should be emptied regularly. Bags of rubbish awaiting collection should not be kept in corridors, stairways or by fire exit doors.

2.5.2.2 Gases and easily vaporisable liquids

All the time that these are contained within the equipment that is designed to hold them, whether it be pipeline, storage tank, reaction vessel, portable container, cylinder, etc., they do not present a fire risk. It is only when they are released to atmosphere either because their use requires it or inadvertently that they become a fire risk. The possibility of inadvertent release can be reduced by careful design of the equipment and training in strict systems of work. The prevention of accidental releases from leaking joints or pipe failure requires high standards of design, installation and maintenance of the equipment.

2.5.2.3 Flammable liquids

These present less of a fire risk than flammable gases because they have higher boiling points and hence require higher ignition temperatures. Flammable liquids should be kept in properly designed containers in a flame-proof store designated for the purpose. Only the minimum amount of flammable liquid should be kept in working areas where it should be kept in a special fire-resistant cabinet or cupboard. Special non-spill self-sealing containers should be provided for the transport and use of flammable liquids in the working area (*Figure 2.4*). Any spillage should be contained by an absorbent and removed as quickly as possible. Flammable-liquid storage tanks should be surrounded by a bund to contain any spillage that may occur.

2.5.2.4 Flammable compressed gases and liquefied gases

Cylinders containing compressed flammable or liquefied gases should be stored in an upright position in the open air and protected from direct

Figure 2.4 Safety can cut away to show stopper. (Courtesy Walter Page (Safeways) Ltd)

sunlight. Oxygen cylinders should never be stored next to flammable-gas cylinders. When in use compressed-gas cylinders should be secured in the upright position.

2.5.2.5 Dusts

Dust is created, not only when solid material is worked on, but also whenever materials are moved or handled. A dust presents a higher fire risk than the solid because of its greater contact area with air with resultant lower ignition characteristics. A dust cloud has many of the fire characteristics of a gas. Accumulations of dust should be removed regularly by techniques which do not generate a dust cloud, i.e. damp sweeping or vacuum cleaning. The electrical equipment in areas where flammable solids and powders are handled or occur must be designed and maintained to the appropriate flame-proof standard.

2.5.3 Products of combustion

When a substance burns it gives rise to four primary products: smoke, products of combustion, light and heat. In the perfect state the process of burning changes the fuel into carbon dioxide, water vapour and the oxides of other elements in the fuel all or nearly all of which are colourless gases. In practice there is incomplete combustion in which not all the fuel is oxidised resulting in the production of carbon and other partially oxidised materials. In general the amount of smoke a fire produces is determined by the amount of air available for combustion: with less air more smoke is produced.

The other products of combustion vary widely in their composition as they clearly depend on the fuel that is burning but they will always contain some carbon monoxide and carbon dioxide as well as water vapour. The hot gas rising from a fire will always be a mixture of the products of combustion, oxygen that has not been consumed in the fire and nitrogen from the air. It is hazardous to health because it is hot, has a lower oxygen content than air and contains toxic substances.

Most fires give off light but often this is obscured by smoke. The light comes from glowing carbon particles and the less yellow the flame the higher its temperature.

Some of the heat that a fire produces is carried from the fire by the smoke and the products of combustion as they rise in the air.

It is the combination of the hazardous nature of the products of combustion together with their high temperature which is the main cause of death in fires.

2.5.4 Fire spread

Four factors are involved in the spread of fire: conduction, convection, radiation and physical transport.

2.5.4.1 Conduction

If a solid body, a metal rod for instance, is heated at one end, the end remote from the source of heat will in time get hot because of the conduction of heat through the rod. The ability to conduct heat varies considerably according to the material, e.g. metal is a much better conductor than wood.

2.5.4.2 Convection

Hot air rises because its density is less than that of the surrounding colder air. The upward movement of hot air is known as convection. The hot air, including smoke, rises until it reaches ceiling level when it begins to spread horizontally. The temperature of the layer of smoke will continue to increase until eventually it will ignite any combustible material with which it comes into contact. If the compartment in which the fire occurs is large enough, the spreading layer of smoke will begin to cool and start sinking, creating a 'mushroom' effect. Any open staircase will act as a chimney and carry smoke and heat to upper floors.

2.5.4.3 Radiation

Heat energy from a hot body radiates as electromagnetic waves which will heat any liquids or solids which they contact. An example of radiated heat are the rays from the sun. Radiated heat can pass through some materials, such as glass, and ignite combustible materials on the other side.

2.5.4.4 Physical transport

Fires can be spread by the collapse of hot or burning materials. They can be spread by the transport of burning debris in the updraught of the fire or by the flow of burning liquids down a slope.

2.5.5 Combustion properties of building materials and construction

Buildings are rarely, if ever, made wholly of one material, so the fire risk on the structure can be complex. Common building materials include wood, steel, concrete, brick, stone, glass, plaster and finishing materials, all of which display different characteristics when exposed to fire. Steel expands when heated, for example a 10 m steel joist will expand by 60 mm when its temperature rises by 500°C and will lose two-thirds of its strength at 600°C. Also at around 600°C the free quartz in granite will expand and shatter the stone. On the other hand, when wood burns its external surface chars and, due to its high thermal insulation properties (poor conduction), the timber behind the

Figure 2.5 Aftermath of fire in textile mill. (Courtesy Huddersfield Examiner)

charred surface is protected and remains undamaged. Thus wood can retain its structural strength much longer than other materials.

Fire in a building will normally spread upwards quickly, much more slowly horizontally and very slowly in a downward direction. Any vertical shaft in a building, such as a stairwell or liftshaft, is an ideal route for fire spread and this is why they need to be well protected and isolated from the rest of the building. *Figure 2.5* illustrates the aftermath of a fire in a textile mill.

A fire in a room with the door closed will take a long time to spread and will often die down or may even go out due to having consumed all the oxygen. This is why, in a fire situation, it is important to close doors whenever possible and to be very careful when opening a closed door since the inrush of fresh air containing oxygen can cause a sudden reignition of the fire with explosive force.

When a fire occurs in a large open-plan area of a building the smoke and heat cannot escape and will build up at ceiling level, then spread outwards possibly igniting other materials. One way of preventing this is to install in the roof vents having fusible links that open automatically allowing the heat and smoke to escape (*Figure 2.6*). A basic fire prevention principle is to divide a building into separate fire-resisting compartments (compartmentisation) having fire-resistant walls, floors, ceiling and doors. The object being to contain a fire within one compartment.

Figure 2.6 Automatic fire ventilators on a roof discharging large quantities of heat and fume, so reducing the risk of fire spread. (Courtesy Colt International Ltd)

2.6 Protection against fire

2.6.1 Structural precautions

The three key elements in structural fire resistance are insulation, integrity and stability. Insulation prevents the passage of heat by conduction through an item of structure such as a wall. Integrity refers to the prevention of the passage of flames and hot gases through the element of structure, while stability is concerned with the ability to resist collapse. As a general rule the minimum fire resistance that is required of a structure or element of structure is 30 minutes but in certain cases it could be several hours.

When the fire resistance of buildings is being considered attention must be given to a range of features that are effective in controlling fire spread.

2.6.1.1 Structural elements

Unless the structural members of the building possess the necessary fire resistance they will require protecting. This can be achieved by enclosing the structural member in concrete or in fire-resistant panels or insulation.

2.6.1.2 Walls, doors and openings

Walls, doors and other openings should be designed to resist the passage of fire. Brick, concrete and other non-combustible wall materials will do this effectively. In corridors and passages doors stop the spread of fire and smoke. Wooden doors can be designed to provide good fire resistance although careful thought must be given to the type and construction of the door where fire resistance is important. To prevent the spread of smoke, doors must be close fitting and there should be a minimum gap between the leaves of double doors. Fire doors should be either self-closing or close automatically when a fire is detected. Other openings in floors and walls, such as for conveyors, should be provided with a suitable fire-resisting door or shutter that will close automatically in the event of a fire.

2.6.1.3 Roofs and floors

Roofs and floors can be affected by fire from below when heat and flames rise, and from a fire above by flame impingement. The outer surfaces of roofs and walls should wherever possible be constructed from non-combustible material.

2.6.1.4 Stairways and liftshafts

Stairways and liftshafts usually communicate with the different levels in a building and can provide an effective path for a fire to spread. The stairways and liftshafts should be enclosed in a fire-proof structure and any doors into them should be fire resisting.

2.6.1.5 Cavities and voids

Cavities and voids are present in most buildings and most occur commonly between the outer and inner walls, above suspended ceilings and underneath floors. They often contain the services for the building. Cavities and voids should be separated by a fire-resistant partition to resist the spread of smoke and heat. In particular where a cavity occurs over a fire-resistant wall, the wall or an equivalent fire-resistant partition should be extended to the ceiling or roof of the cavity.

2.6.1.6 Cables

Buildings contain electricity cables which commonly have PVC insulation. When exposed to fire this insulation will burn giving off smoke and toxic gases. Cables which serve a key service such as emergency lighting, fire alarm systems and fire water pumps should be fire resisting or of the mineral insulated type. It is possible for the cable itself to be the source of fire through overheating because it is overloaded. If cables are run in a service way or duct, heat and/or smoke detectors and automatic firefighting devices can be installed to protect them.

2.6.1.7 Compartmentisation

While open-plan buildings and stores have benefits from a business operating point of view, they do nothing to prevent the spread of smoke and flames in the event of fire. Wherever possible large spaces in buildings should be divided into smaller compartments by fire-resisting walls which should be carried up to the ceiling or roof. Any openings in these walls, for doors, conveyor ways etc., should be provided with automatic fire-resisting doors or shutters.

2.6.1.8 Smoke control

Smoke is not only hazardous to health, it is also hot and where it accumulates its heat can be passed to the surrounding materials thus extending the fire. By placing ventilators in the roof of large open-plan buildings the build-up of smoke and heat can be prevented. An additional benefit of these ventilators is that they create a chimney effect over the fire which draws the flames and smoke upwards reducing the lateral spread of the fire (*Figure 2.6*).

2.6.1.9 Plant

Buildings may contain plant which needs special fire protection because of its nature. At one extreme these include storage tanks containing liquefied flammable gases which may require water drenching for cooling, while at the other extreme, cupboards containing vital records should be of metal construction with effective heat insulation on the inner surface.

2.6.2 Fire alarms and detectors

2.6.2.1 Fire alarms

Alarms may be manual or automatic. Manual alarms are hand-operated devices such as gongs or bells which are only suitable for small premises. The most common manually operated electric alarms, which are preferred and should be considered as standard wherever possible, comprise break glass call points which, when operated, sound an alarm. Normally the alarm will be an audible bell, siren or hooter which must be loud enough to be heard in all areas frequented by employees, including the toilets. The alarm could also be a distinctive sound or message. In some places of public assembly, such as cinemas and sports stadia, it could be in a coded form which is recognised by suitably trained members of staff who then pass the message to the public.

Automatic alarms operate by detecting particular changes in the environment. The change detected will depend on the type of detector but can be identified with the stages of development of a fire:

1 The presence of invisible products of combustion that are released soon after ignition occurs.
2 Visible smoke for the burning material.
3 Flames and a degree of illumination.
4 The temperature in one part of the protected area rising rapidly or rising above a predetermined level.

Automatic fire alarms have the advantage of being able to raise the alarm in the event of a fire in an unoccupied or unmanned area. The stage at which a fire is detected depends on the detector of which there are three main types:

(a) smoke detectors
(b) heat detectors
(c) flame detectors.

2.6.2.2 Smoke detectors

Smoke is a complex mixture of gases, liquids and solid particles depending on the materials being burnt and the surrounding conditions. Each of the constituents of smoke displays particular optical and physical properties which are made use of in the two main types of detectors – optical and ionisation.

Optical detectors work by detecting the obscuration and/or scattering of a focused light beam by the smoke particles (*Figures 2.7* and *2.8*). Beam-type optical detectors can be used to protect open working spaces by projecting a beam across the space to a receiver at the far side; any smoke interferes with the beam of light and causes the detector (receiver) to trip the alarm. This type of detector is also used to protect warehouses.

Ionising detectors sense the electric current generated in the instrument by the radioactivity and react to any changes in that current that result from the smoke interfering with the radioactive emissions (*Figure 2.9*).

2.6.2.3 Heat detectors

Heat detectors are of either the fixed temperature type or rate of temperature rise type. Fixed temperature detectors are activated when the temperature in the area reaches a predetermined level. Rate of temperature rise detectors are activated when there is an abnormally rapid increase in the temperature at the instrument. They can incorporate

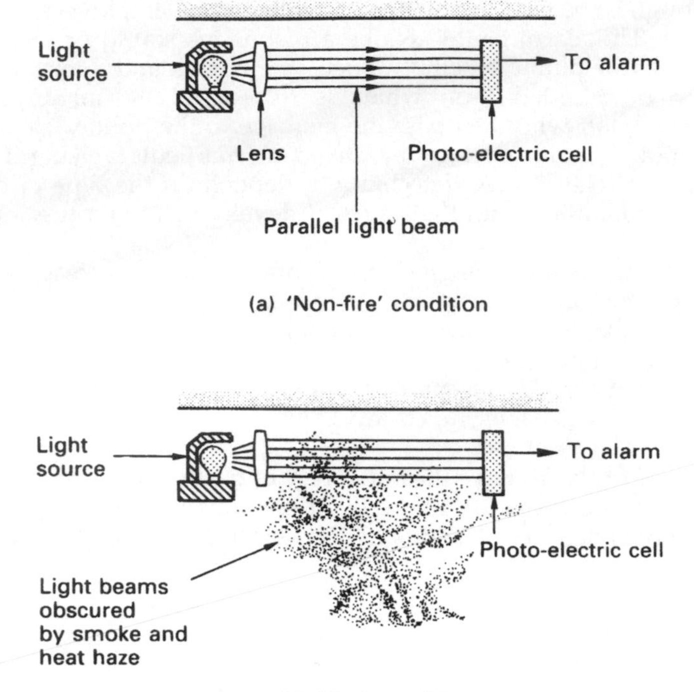

Figure 2.7 Combined heat and smoke detector. (Manual of Firemanship. Courtesy Controller HMSO)

an upper temperature setting. Heat detectors are most effective in areas where there may be smoke or steam under normal conditions, i.e. boiler rooms, kitchens or areas where, in the event of a fire, it can be expected to be a flaming fire with little or no smoke.

2.6.2.4 Flame detectors

Flame detectors are designed to detect either the infrared or the ultraviolet radiation emitted by the flame of a fire. Sunlight, welding or ultraviolet lamps can cause false alarms in this type of detector. Flame detectors are often installed in conjunction with smoke and heat detectors.

2.7 Extinction

Once a fire has started the control of its development and its ultimate extinction can be achieved by:

- reducing the amount of fuel available for combustion (fuel starvation),

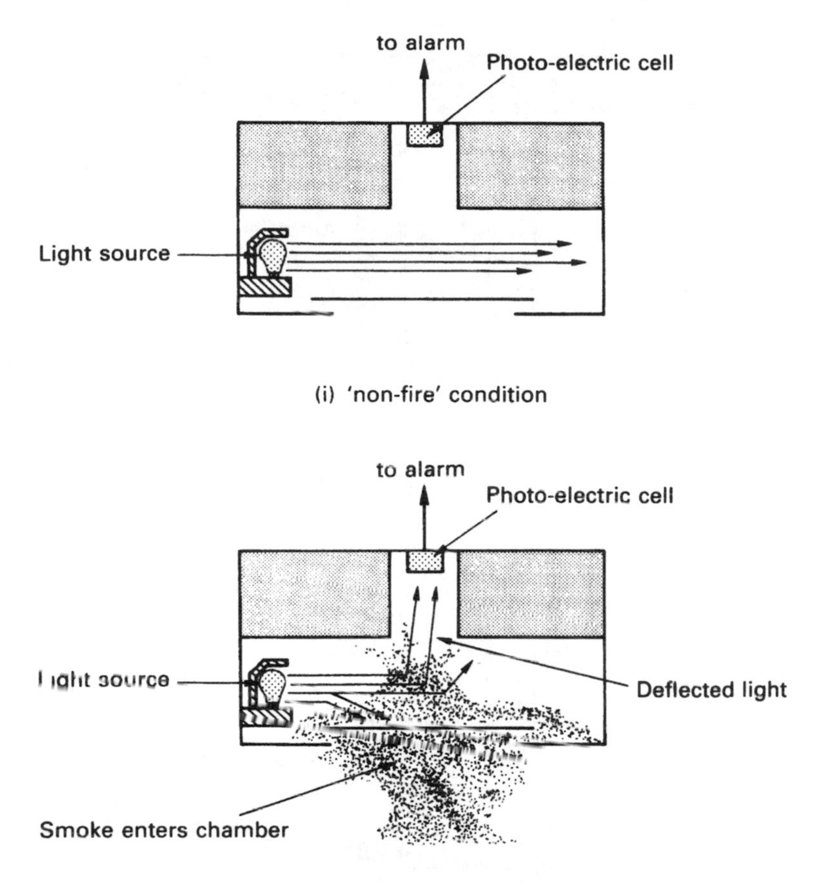

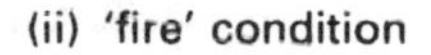
(ii) 'fire' condition

Figure 2.8 Diagram of smoke detector – optical light scatter type. (Manual of Firemanship. Courtesy Controller HMSO)

- reducing the amount of oxygen available for combustion (smothering),
- removing the heat of the fire (cooling),
- direct interference with the oxidation reaction.

Often more than one of these features is employed in fighting a fire at any given time.

2.7.1 Fuel starvation

Fires can be starved of fuel in a number of ways:

- By removing combustible materials from around the fire, e.g. by removing unburnt stock from a fire in a warehouse.

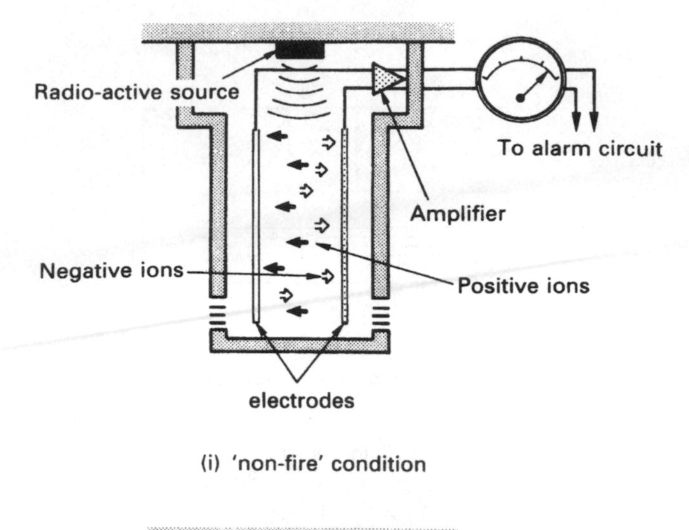

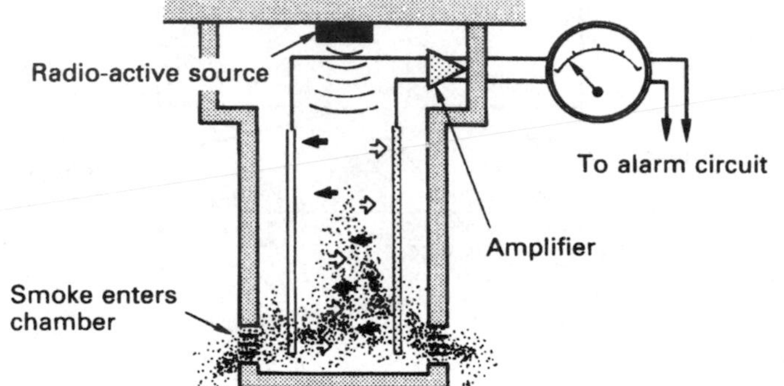

Figure 2.9 Diagram of smoke detector – ionisation type. (Manual of Firemanship. Courtesy Controller HMSO)

- By removing the burning materials from the surrounding unburnt materials, e.g. cutting out a burnt area from a fire in a thatched roof.
- By dividing the burning materials into smaller fires that can be extinguished more easily or be left to burn out, e.g. beating out a heath fire.
- In gas fires, by turning off the gas supply.

2.7.2 Smothering

Where a pool of flammable liquid is on fire, the oxygen supply can be reduced by covering the surface of the liquid with a layer of foam. On the other hand, a fire in a confined space can be extinguished by filling the space with an inert gas and so reducing the oxygen available for combustion. The same effect can be achieved by closing the doors and windows of a room which contains a fire and allowing the fire to consume the oxygen and hence self-extinguish. However, it is important to remember when extinguishing by oxygen removal: if the supply of oxygen is restored or the supply of inert gas fails then the fire may reignite if the heat source is still present. The technique of oxygen removal is of no use if the supply of oxygen comes from the substance that is burning, e.g. ammonium nitrate or a fire involving an oxidising agent.

2.7.3 Cooling

This is the commonest way of extinguishing a fire. A substance is put into the fire which readily absorbs heat thus preventing the fire itself from heating the unburnt fuel up to ignition temperature. Water is the most common heat removal agent because it is cheap, readily available and has a high latent heat of vaporisation. With large fires, heavy jets are needed to provide the necessary quantity of water (*Figure 2.10*).

2.7.4 Interference with the reaction

Fire is propagated by means of a free radical chain reaction and so anything which will interrupt this chain will cause the fire to die. The

Figure 2.10 Use of large quantities of water to quench fire. (Courtesy Liverpool Daily Post and Echo PLC)

'halons' family of chemicals, which are chloro-, bromo-, fluoro-hydro-carbons, are very efficient at this but can have undesirable side effects in that their breakdown products can be toxic. Other safer extinguishants are being developed. Spreading the heat over large surface areas, as in flame traps or with the powder of a dry powder extinguisher, also effectively removes the free radicals and hence results in a suppression of the fire.

2.8 Firefighting

All workplaces must be provided with adequate and appropriate means for fighting fires, the location of which must be suitably indicated and be so placed that it is readily available at all times. The equipment can be portable or fixed.

2.8.1 Portable fire extinguishers

Portable firefighting equipment comprises mainly fire extinguishers although fire blankets and buckets of sand are also included. Fire extinguishers purchased after January 1997 should comply with BS EN 3[1] which requires the bodies of all fire extinguishers to be painted red although BS 7863[2] permits manufacturers to provide a coloured panel in line with the colour coding system to indicate its contents which has been used in the UK for many years, provided that the panel (or band) does not cover more than 5% of the extinguisher body area. The colour coding system divides extinguishers into five main groups according to the extinguishant they contain:

Extinguishant	*Colour code*
Water	Red
Foam	Cream
Carbon dioxide	Black
Dry powder	Blue
Vaporising liquids	Green

Each extinguisher should be marked with numbers and letters, such as 13A, 34B etc., to indicate the maximum size and type of fire that they are capable of extinguishing. These ratings are explained in part 3 of BS 5306[3].

The classification of the types of fires for which each extinguisher is appropriate is given in *Table 2.3*.

2.8.1.1 Operation of portable extinguishers

Modern extinguishers work by an internal pressure causing the ejection of the contained extinguisher medium through an operating lever and

Table 2.3 Classification of fires for portable extinguishers

Class of fire	*Description*	*Appropriate extinguisher*
A	Solid materials, usually organic, with glowing embers	Water, foam, dry powder, vaporising liquid and CO_2
B	Liquids and liquefiable solids:	
	(i) miscible with water, e.g. methanol, acetone	Water, foam (but must be stable on miscible solvents), CO_2, dry powder, vaporising liquid
	(ii) immiscible with water, e.g. petrol, benzene, fats and waxes	Foam, dry powder, vaporising liquid, CO_2
C	Gas or liquefied gas	–
D	Metals	–
–	Electrical equipment	Dry powder, vaporising liquids, CO_2

Note: Particular dry powders may be required for use on different classes of fires.

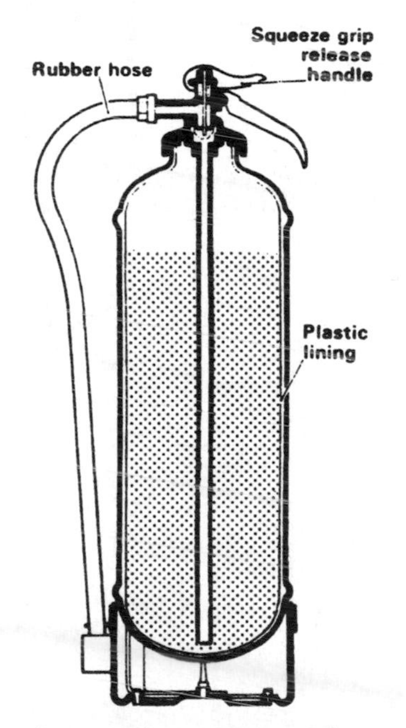

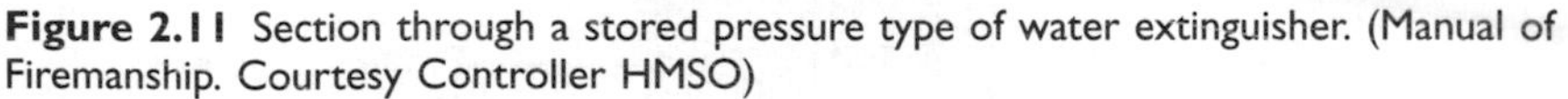

Figure 2.11 Section through a stored pressure type of water extinguisher. (Manual of Firemanship. Courtesy Controller HMSO)

discharge pipe with nozzle. The extinguisher cylinder is either pressurised – stored water type (*Figure 2.11*) and carbon dioxide (*Figure 2.12*) – or contains a high-pressure cartridge which has be pierced to cause pressurisation of the cylinder – water type (*Figure 2.13*), foam extinguisher (*Figure 2.14*) and dry powder extinguisher (*Figure 2.15*). The

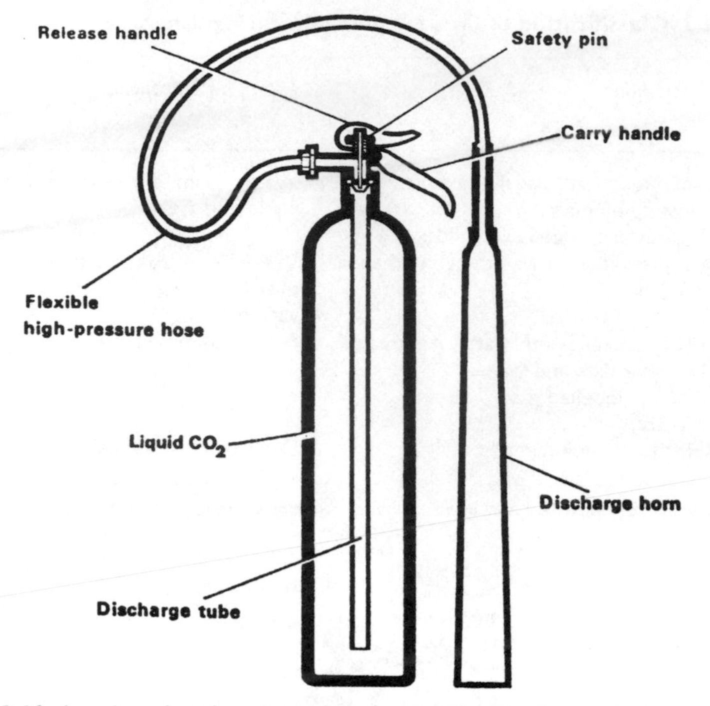

Figure 2.12 A carbon dioxide extinguisher showing the piercing mechanism, control valve and discharge horn. (Manual of Firemanship. Courtesy Controller HMSO)

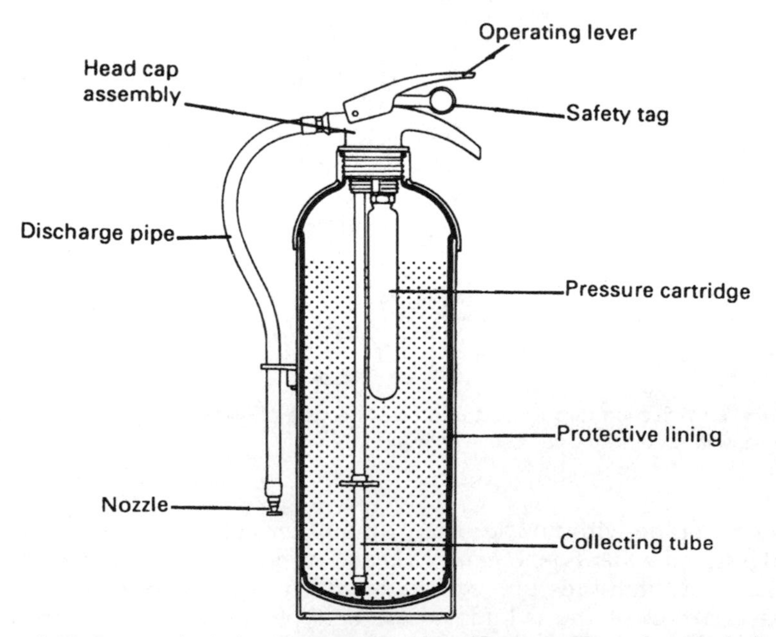

Figure 2.13 Section through a water extinguisher. (Courtesy Thorn Security Ltd)

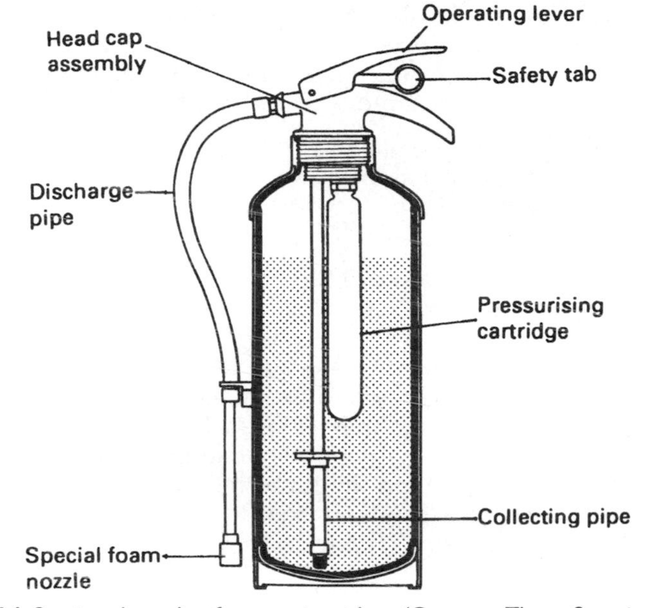

Figure 2.14 Section through a foam extinguisher. (Courtesy Thorn Security Ltd)

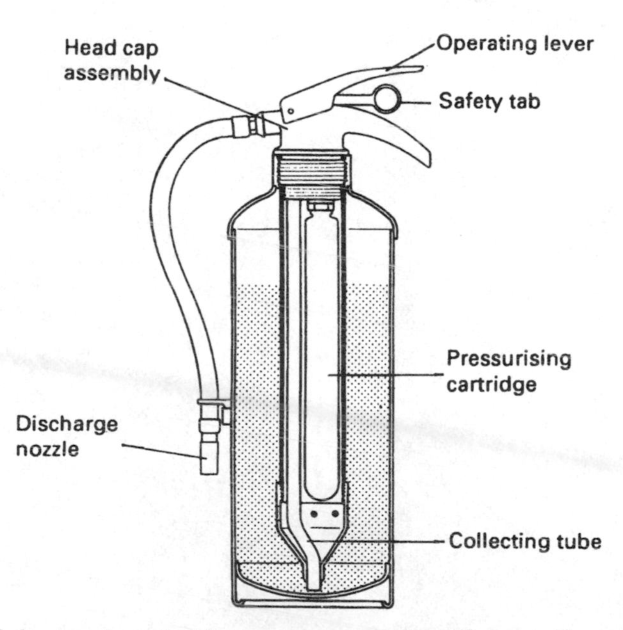

Figure 2.15 Section through a dry powder extinguisher. (Courtesy Thorn Security Ltd)

operating lever allows control of the flow of extinguishing medium while the discharge pipe allows the flow to be directed as required. Most extinguishers have a safety tag incorporated in the operating lever to prevent its inadvertent discharge.

2.8.1.2 Provision, location and maintenance of portable firefighting equipment

(a) Provision

Portable fire extinguishers are of little use unless of the correct type, in sufficient numbers and placed so as to be readily available at all times. Staff must be trained in their use and must also know which extinguisher is most suitable for the different classes of fires. Nobody should have to travel more than 30 metres in any direction to obtain a fire extinguisher. Extinguishers should be placed in positions where they are clearly visible at all times and where they will not be obscured by storage of goods or other equipment. Their position should be indicated by a notice and they should be mounted on a coloured board, with outlines so it will be obvious when one is missing (*Figure 2.16*).

Figure 2.16 Clear unobstructed fire point. (Courtesy Reed Medway Sacks Ltd)

(b) Location
Extinguishers should be located:

- On exit routes and wherever possible adjacent to exit doorways.
- If possible at the same location on each floor level.
- If practicable, sited with fire alarm call points and fire action notices to form a fire point.
- Away from extremes of temperature.

(c) Maintenance
Fire extinguishers require to be inspected and maintained regularly. Monthly routine inspection should be carried out in-house to ensure that extinguishers are correctly sited, have suffered no obvious damage and not been discharged. Where plastic tags, which have to be broken before an extinguisher can be operated, have been fitted the monthly check should ensure they have not been broken. Annually, extinguishers should be stripped down and given a thorough inspection by a competent and suitably trained person. The date of this inspection should be recorded on the extinguisher. Every 5–10 years the extinguisher should be discharged. Full details of testing and maintenance are contained in part 3 of BS 5306[3].

2.8.2 Fixed firefighting equipment

These can be split into two categories:

- manually operated such as hose reels and rising mains for use by the fire brigade
- automatically operated such as sprinklers and drenchers.

2.8.2.1 Manually operated systems

(a) Hose reels
Hose reels usually comprise about 30 m of hose wound onto a metal drum and connected permanently to the mains water supply. Their location should ensure that no part of the building is more than 6 m from the nozzle when the hose is fully run out. Water pressure should be sufficient to produce a 6 m jet of water from the highest hose reel. Hose reels have the advantage over extinguishers in that they can deliver an unlimited supply of water yet still be easy to operate and the nozzles can be adjusted to change from a jet to a spray or shut down completely at will. They should be checked monthly to ensure there are no leaks and that the outlet nozzle is not blocked. Once a year all the hose on the drum should be completely run out and the hose inspected and given an operational test.

(b) Rising main
There are two types of rising main, wet and dry. They are of the same construction with a pipe passing up through the building and having a

landing valve at each floor level. The landing valve allows the fire brigade to attach their hoses directly to it, saving time in getting water to the seat of the fire. As its name implies, the wet rising main is kept permanently charged with water either from a tank or by a booster pump. A dry rising main is empty with an inlet at its lowest point to which the fire brigade can connect their pump and so charge the main.

2.8.2.2 Automatically operated systems

2.8.2.2.1 Automatic sprinkler systems

Sprinkler systems are designed to provide an automatic means for detecting and extinguishing or controlling a fire in its early stages. Although their role in the past has been primarily for the protection of buildings they are now having an increasing role in life safety. A sprinkler

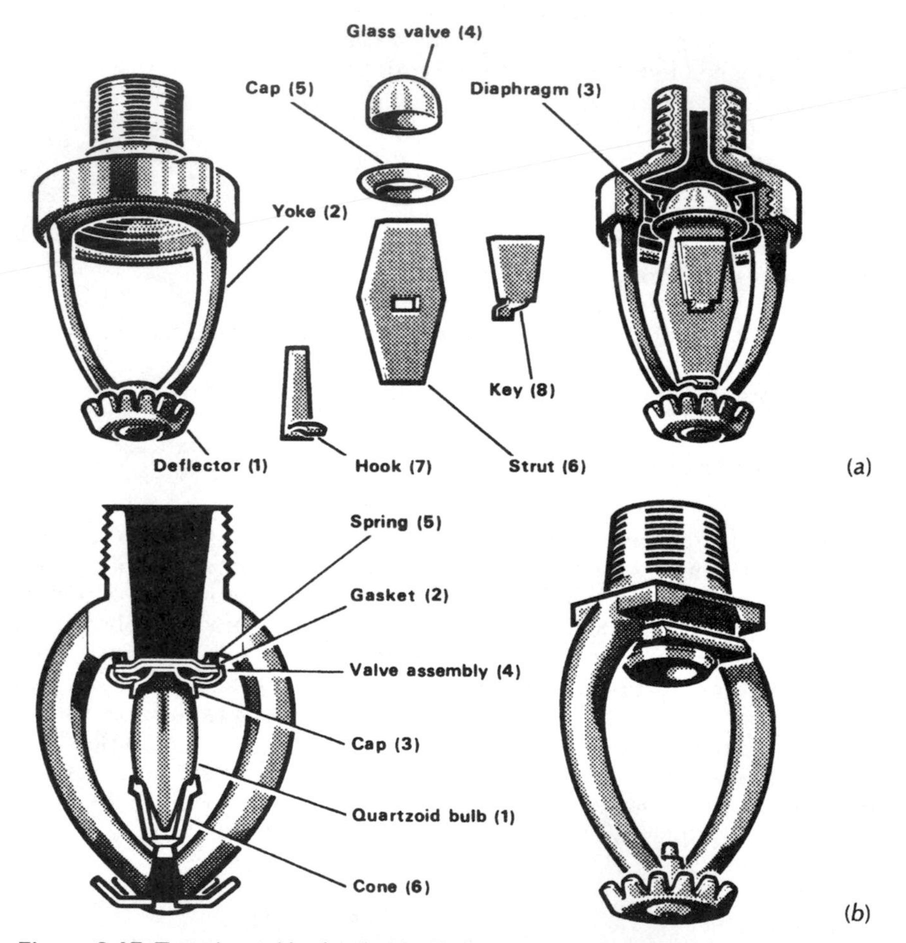

Figure 2.17 Typical sprinkler heads (a) a fusible solder type sprinkler head; (b) bulb type sprinkler head. (Manual of Firemanship. Courtesy Controller HMSO)

system comprises a range of pipework with regularly spaced sprinkler heads. The sprinkler heads are activated by heat when they discharge water onto the fire, and at the same time the drop in water pressure in the main trips the alarm.

A sprinkler head is an automatic water valve fitted with a deflector plate to give a specific spray pattern. The valve is sealed by a heat sensitive element which is designed to burst at a predetermined temperature. The heat sensitive element is either a specially designed solder seal or a heat sensitive bulb (*Figure 2.17*). The temperature at which the water is released is indicated by a colour code on the yoke arm for the solder type and by the colour of the liquid in the bulb of the bulb type (*Table 2.4*). One advantage of a sprinkler system is that only the heads in the vicinity of the fire operate, thereby reducing the extent of the water damage.

Four types of sprinkler system are available: wet pipe, dry pipe, alternate and pre-action. A *wet pipe system* is the most common where the pipework is filled with water under pressure. When a sprinkler head operates, water immediately discharges through it and continues until the system is turned off. This system is only suitable for buildings where it will not be subject to frost.

A *dry pipe system* can be used in unheated areas where water in the piping could be susceptible to freezing. Instead of water, the pipework contains air at a pressure which is sufficient to hold a head of water behind a valve which is located in an area not subject to frosts. When a sprinkler head operates, the drop in the air pressure allows that valve to lift releasing water into the pipework and hence to the particular head that has operated. A dry pipe is less efficient than a wet pipe system since there is a time delay before the water is discharged onto the fire. This time delay can be reduced by the use of accelerator valves that release the contained air.

Table 2.4. Sprinkler head colour codes

Soldered type	°C	*Colour of yoke arm*
	68–74	Uncoloured
	93–100	White
	141	Blue
	182	Yellow
	227	Red
Bulb type	°C	*Colour of bulb liquid*
	57	Orange
	68	Red
	79	Yellow
	93	Green
	141	Blue
	182	Mauve
	204–260	Black

An *alternate system*, which is sometimes used in areas that are subject to freezing only during cold weather, allows either wet or dry operation to be selected.

A *pre-action system* is similar to a dry pipe system but is considered to be more efficient. It is used where an alarm is required in advance of the operation of the sprinkler heads. This is particularly useful in preventing accidental operation of the system. In a pre-action system heat or smoke detectors are installed throughout the protected area. When a detector is actuated an automatic water control valve admits water into the sprinkler pipework where it remains at pressure until a sprinkler head operates.

2.8.2.2.2 Drencher systems

Ensuring adequate firefighting provisions where large quantities of liquid or gaseous flammable materials are handled, e.g. chemical plants, tank farms, loading and off-loading installations, requires very large quantities of water. When a fire occurs in these situations it is often far more important to provide adequate cooling of the installation than it is to extinguish the fire. The vast amounts of water required call for excellent supply arrangements and when used can cause extensive flooding on the site if adequate arrangements have not been made to contain or direct the disposal of the fire water. It should be remembered that fire water is often contaminated by chemicals and its release into natural water courses should be prevented since it can cause severe environmental damage.

A drencher system operates in a similar manner to the pre-action sprinkler system except that the sprinkler heads are open (they have no heat sensitive element) so that on actuation of a separate detector, water is pumped into the pipework and is discharged through all the sprinkler heads.

2.8.2.2.3 Fixed foam installations

Foam installations are provided where there is a risk of flammable liquid fires, for confined spaces where access is difficult and for the total flooding of high-risk spaces. The systems can be automatically operated and may be connected to detector and alarm systems.

2.8.2.2.4 Carbon dioxide systems

Carbon dioxide provides good protection for hazardous plants such as electrical transformers and switchgear and also for computer rooms and control rooms. Carbon dioxide is usually stored as a liquid in large cylinders and connected by a system of pipework to discharge nozzles located in the protected area. Systems can be automatic but usually have a facility to be switched to manual operation in occupied areas. Since carbon dioxide is an asphyxiant, if the system is automatically operated a time delay and alarm must be built in to allow any occupants of the room to escape before carbon dioxide is discharged.

2.8.2.2.5 *Halon systems*

Halon systems are very similar to those for carbon dioxide except that for a similar risk the storage cylinders would be smaller due the differences in extinguishing action. However, following the Montreal Protocol, halon production in the UK ceased at the end of 1993 although existing installations can continue to be used while the supply of recycled halon lasts. Alternate inert gas products are being developed and will become available.

2.9 Fire risk

There are three categories of fire risk applicable to places of work: high, normal and low. The factors which lead to a workplace or parts of it being assessed as being 'high risk' include:

(a) whether the workplace is used for sleeping accommodation;
(b) whether the materials present are either easily ignited or, when alight, are likely to cause the rapid spread of fire, smoke or fumes;
(c) the presence of unsatisfactory structural features;
(d) whether there are certain parts of the workplace which could present a greater risk of fires occurring and developing than elsewhere; and
(e) any special consideration which may have to be given to occupants, either those who work in the workplace or others present.

Most workplaces will fall within the 'normal risk' category where the factors to be considered are:

(f) any outbreak of fire is likely to remain localised or is likely to spread only slowly;
(g) there is little risk of any part of the structure or contents of the building taking fire so readily or producing large quantities of smoke or fumes as to be a threat to life safety; and
(h) the nature of the occupancy does not require a higher risk classification.

A workplace may be assessed as of 'low risk' where the risk of fire occurring or of fire, smoke or fumes spreading is negligible. Such a workplace may be a heavy engineering shop or one where the process is entirely a wet one and non-combustible materials predominate.

2.10 Means of escape in case of fire

The principles on which means of escape provisions are based are that persons, regardless of the location of the fire, should be able to proceed safely along a recognised escape route, by their own unaided efforts, to a place of safety. Separate considerations apply to disabled persons. The characteristics of an escape route include:

(a) there should be an alternative means of escape from most situations;
(b) the distance that a person should travel to a place of safety is dependent upon the risk assessment of the workplace;
(c) where direct escape to a final exit is not possible, a place of relative safety, such as a protected escape route, should be reached within a reasonable travel distance;
(d) unless a place of safety can be reached within a reasonable travel distance the escape routes will need to be protected from the effects of fire;
(e) escape routes should be wide enough to cater for the number of occupants, e.g. not less than 1.05 m in width, and should not reduce in width; and
(f) from every room, storey or building there should be a sufficient number of available exits of adequate width.

The following are not acceptable as means of escape in case of fire: lifts (except where specially designated), escalators, portable or similar ladders, self rescue apparatus and lowering devices.

2.10.1 Means of escape for disabled people

Means of escape standards are based on the assumption that all occupants are able-bodied. Approved Document M1[4] of the Building Regulations 1991 gives detailed advice and guidance on the facilities that must be provided for the disabled. Part 8 of BS 5588[5] lays down standards which fire officers will apply when assessing any additional facilities that may be necessary for the safety of disabled people.

2.10.2 Travel distance

Suitable travel distances, in metres, are given in *Table 2.5*.

2.10.3 Exits

More than one exit is required when the number of occupants of any room exceeds 60, when the travel distance to the only exit exceeds the value given in *Table 2.5* or when the room contains a high fire risk.

The minimum width for an exit should be 750 mm and where more than one exit is provided they should be more than 45° apart when viewed from the farthest point in the room.

Table 2.5 Escape travel distances (in metres)

Type of premises	*Factory*			*Shop*		*Office*
Category of risk	high	normal	low	high	low	normal
Within a room or enclosure						
One direction only	6	12	25	6	12	12
More than one direction	12	25	50	12	24	25
From any point in an inner room to the exit from the access room	6	12	25	6	12	12
Total travel distance to storey exit (see note (a))						
One direction only	12	25	45	12	18	25
More than one direction	25	45	60	25	30	45

Note (a): *Storey exit* is an exit which, once through, persons are no longer at immediate risk from the effects of fire and includes a final exit, an exit to a protected lobby or protected stairway including an external stairway.

2.10.4 From the site of fire to a place of safety

(a) Gangways and passages

Gangways and passageways in rooms from which persons need to escape in case of fire should be so arranged and kept clear so that there is a free passage at all times to the exit.

(b) Corridors

Corridors which form part of the escape route should be kept clear and unobstructed throughout their length. Where escape is in one direction only, the route should be a protected route, i.e. protected from fire in adjoining accommodation by fire-resistant construction. Corridors should be at least 1.05 m wide and, where in shops they exceed 30 m in length, and in offices and factories 45 m, they should be subdivided by fire doors to prevent the free travel of smoke and other products of combustion.

(c) Stairways

It is good practice to provide at least two stairways in any building. Stairways should be at least 1.05 m wide. All stairways should be separated from the remainder of the building by fire-resisting construction and fire doors such that the stairwell is enclosed.

(d) Doors

Doors to and on the escape route should open in the direction of travel, each leaf should open not less than 90° and when open they should not obstruct any escape route. Fire doors protecting an escape route should be fitted with a self-closing device, smoke seals and provide 30 minutes' fire resistance. Devices to retain the doors open are permitted provided they allow the door to close automatically in the event of a fire. Doors used as a means of escape must be kept unlocked at all times when people are in the building. Panic bolts and latches are permitted to ensure security but in no case should a door be fastened in such a way that it cannot be opened easily and immediately from the inside without the use of a key.

2.10.5 Signs and notices

All fire safety signs and graphic symbols should comply with the Health and Safety (Safety Signs and Signals) Regulations 1996 and with BSs 5378[6] and 5499[7]. These include fire exit and directional signs, signs on doors such as *'Fire door keep shut'* and *'Push bar to open'* and also signs that indicate the location of firefighting equipment and fire alarm call points.

25.10.6 Escape route lighting

All escape routes, including both internal and external stairways, should be provided with artificial lighting of sufficient intensity to enable persons to see their way out of the building safely when there is insufficient natural light. Switches to control the lighting should be provided at the entry point into the escape route and, where necessary, should be clearly indicated. In parts of a building where there is no natural lighting, such as underground areas, a system of emergency escape lighting should be provided which will illuminate automatically in the event of a failure of the main electrical supply.

2.10.7 Emergency plans

The aim of an emergency plan is to ensure that, in the event of a fire, all employees are sufficiently familiar with escape procedure and the fire safety arrangements to take the correct action and evacuate the premises safely. The plan should specify the fire responsibilities of the managers and other nominated responsible persons. It should also detail the actions for:

(a) operating the fire warning system;
(b) calling the fire brigade;
(c) evacuating the occupants to predetermined assembly points in a safe place;
(d) attacking the fire if it is safe to do so;
(e) stopping machinery and plant and isolating power supplies where appropriate;
(f) closing doors; and
(g) liaising with the fire brigade on arrival.

When the fire brigade arrives on site they should be told: the location and extent of the fire, whether any persons are trapped or missing, any known special risks and the location of fire hydrants or other water supplies. The fire brigade should be encouraged to make regular visits to the site to familiarise themselves with the layout of premises and the risks posed.

The evacuation procedure should follow a pre-arranged plan which is regularly rehearsed and updated. It should identify the assembly points and the arrangements for checking that all employees and others on the site are accounted for. Annual fire drills should be carried out to test the emergency procedure and ensure that employees know what to do. Following a fire evacuation of a building, it should not be re-entered without the permission of the fire brigade.

A simple fire instruction, should be displayed at suitable locations in the workplace and typically contain the following information:

Fire Instructions

If you discover a fire –

1 Raise the alarm.
2 Attack the fire using the fire extinguishers provided but not putting yourself at risk.

On hearing the alarm –

3 Leave the building by the shortest route.
4 Go to your assembly area which is...

Do NOT stop to collect belongings.

Do NOT re-enter the building until told to do so.

This notice must be rectangular with white lettering on a blue background as it is a mandatory notice (see BS 5378).

2.10.8 Instruction and training

No matter how good the provisions for fire prevention and protection it will not be effective if the people concerned are not given the necessary instruction and training. Every employee should be given instruction, training and refresher training in:

(a) the means of escape in case of fire;
(b) the action to be taken in case of fire;

(c) the location and method of operating equipment provided for fighting fires; and
(d) the location and use of fire alarms.

Fire marshals and other persons identified in the emergency plan must receive additional instruction and training appropriate to their responsibilities. This should include the checking of workplaces to ensure that they have been evacuated, the carrying out of roll calls at the assembly points and liaison with the fire brigade.

2.10.9 Maintenance and record keeping

All firefighting equipment, fire detectors and alarms must be maintained regularly and kept in efficient working order. Records must be kept of such maintenance which should include details of any work carried out on the equipment. Similarly, records must be kept of all instruction, training and refresher training given to employees, and of all fire drills. The latter should include recommendations to remedy any defects found.

All records must be retained for a period of at least three years from the date on which they were made.

2.11 Legal requirements

Like all areas of occupational activities, fire has its legal requirements, both of a detailed and a general nature.

2.11.1 The Fire Services Act 1971

This places on fire authorities a number of duties including that of securing the services of a fire brigade and such equipment as may be necessary to meet all normal circumstances, the training of the firefighters and the provision of fire safety advice. It also details the powers of the firefighters.

2.11.2 The Fire Precautions Act 1971

The specific requirements contained in the Fire Precautions Act 1971 (FPA) have been considerably modified and extended by the Fire Safety and Safety of Places of Sport Act 1987 (the 1987 Act) and it is necessary to refer to the two Acts when considering an employer's fire obligations.

No specific requirements in respect of fire, fire prevention, or fire precautions are contained in the HSW Act although under s. 78 the Fire Precautions Act 1971 was extended to include places of work. This Act took over the fire clauses of the Factories Act, and Offices, Shops and Railway Premises Act, and introduced the requirement for a fire certificate rather than a means-of-escape certificate. The Act gives power to the Minister to designate premises that require a fire certificate.

2.11.2.1 Premises that can be designated

Under s.1 of FPA six classes of premises can be designated:

(a) Premises used as, or for any purpose involving, the provision of sleeping accommodation.
(b) Premises used as, or as part of, an institution providing treatment or care.
(c) Premises used for the purpose of entertainment, recreation or instruction or for the purposes of any club, society or association.
(d) Premises used for the purposes of teaching, training or research.
(e) Premises used for any purposes involving access to the premises by members of the public, whether on payment or otherwise.
(f) Premises used as a place of work.

The only class of premises that cannot be designated is a single private dwelling. Only two Designating Orders have been made relating to:

(a) Hotels and boarding houses.
(b) Factories, offices, shops and railway premises.

In the case of hotels and boarding houses, a fire certificate is required if sleeping accommodation is provided above the first floor or below the ground floor. For factories, offices, shops and railway premises a fire certificate is required if more than 20 persons are at work in the building or more than 10 work elsewhere than the ground floor. A fire certificate can also be required for factories using or storing highly flammable or explosive materials regardless of how many are employed. The fire brigade have the power to exempt from the requirement to have a fire certificate certain premises that are considered to be low risk.

2.11.2.2 Application for a fire certificate

This has to be made on Form FP1 (Rev) which is available from the local fire authority. Before issuing a certificate the fire authority may inspect the building, ask for more information or plans and ask for remedial work to be carried out.

2.11.2.3 Interim duties

Once the application has been made and prior to the issue of the fire certificate the occupier of the premises must:

(a) Ensure that the existing means of escape are kept clear and usable at all times.
(b) Maintain all existing firefighting equipment.
(c) Train all staff in the actions to be taken in the event of a fire.

2.11.2.4 Issue of a fire certificate

When the fire authority are satisfied with the fire safety standards, they will issue a fire certificate which will detail:

(a) The use or uses of the premises it covers.
(b) The means of escape in case of a fire, usually indicated by way of a plan of the building.
(c) The measures for ensuring that the means of escape can be safely and effectively used at all times, i.e. fire doors, emergency lighting and fire exit signs.
(d) The means for fighting fires for use by persons in the building.
(e) The means for raising the alarm.
(f) In the case of factory premises, particulars of any highly flammable or explosive substances stored or used on the premises.

The fire certificate may also impose requirements concerning:

(g) The maintenance of the means of escape and keeping it free from obstruction.
(h) The maintenance of the firefighting equipment and the fire alarm system.
(i) The training of staff in what to do in the event of a fire and the keeping of records of such training.
(j) Limiting the number of people who may be in the premises at any one time.
(k) Any other precautions they consider relevant.

Once the fire certificate has been issued it must be kept on the premises to which it relates. Any proposed alterations to the buildings that could affect any item on the fire certificate must be notified to the fire authority before the work commences.

2.11.2.5 Appeals and offences

If an applicant for a fire certificate considers any requirement imposed by the fire authority is unreasonable, an appeal can be made to a magistrate within 21 days of being made aware of that requirement.

It is an offence to contravene any requirement imposed by a fire certificate or to use a premises for a designated use without applying for a fire certificate.

2.11.2.6 Powers of a fire authority

If a fire authority are of the opinion that the use of a premises places people at such a serious risk from fire that the use ought to be prohibited or restricted, it can issue a Prohibition Notice. A Prohibition Notice can be issued to any premises covered by s. 1 of the Act whether or not it has been designated. The notice takes immediate effect and will stay in force until the risk has been reduced to an acceptable level. An occupier can appeal to a Magistrates' Court against the issue of a notice, but it stays in force until the decision of the court is known.

An inspector appointed by the fire authority under the FPA has the power:

(a) To enter and inspect at any reasonable time any premises to which the Act applies or appears to apply.
(b) To make such enquiry as is necessary to ensure compliance with the Act.
(c) To require production of and inspect the fire certificate.
(d) To require the owner, occupier etc. to render any assistance he may need in the carrying out of his investigation or enquiries.

2.11.2.7 Charges for fire certificates

The fire authority can make a charge for the issue or amendment of a fire certificate. The charge must relate to the amount of work involved in the preparation of the certificate and not the inspection of the premises.

2.11.3 The Fire Certificates (Special Premises) Regulations 1996

Special premises come within certain categories listed in the Regulations but all are considered to be a high fire risk. In such cases, the Regulations transfer the responsibility for issuing fire certificates from the fire authority to the HSE.

2.11.4 The Fire Precautions (Workplace) Regulations 1997

These Regulations bring into UK law the fire requirements of two EU directives[8,9]. They do not apply to premises that have a current fire certificate issued under FPA. However, they make minor amendments to the Management of Health and Safety at Work Regulations 1992 and introduce the concept of risk assessment into fire legislation for the first time.

2.11.4.1 Requirements of the Regulations

The Regulations require the occupiers or owners of all premises to which they apply to carry out an assessment of the fire risk and to provide adequate fire safety measures against the risks identified. The fire authority will use the fire risk assessment as a starting point when carrying out their inspections. Once the risk assessment has been completed, the provisions made must include:

(a) Adequate means of escape in case of a fire.
(b) Adequate means for giving warning in the event of a fire.
(c) The provision of adequate firefighting equipment.
(d) The preparation of an emergency plan, identification of those who will implement the plan and training for them.

2.12 Liaison with the fire brigade

For the fire brigade to be effective, it needs to get to the source of the fire as quickly as possible. In private dwellings and small offices and workshops, having arrived at the premises, the source of the fire can be easily located. However, with large office blocks and factories and also shopping, leisure and recreational complexes because of their size and layout, the delay in the fire brigade finding their way around the site may be crucial to the successful extinguishing of the fire. In such premises it is prudent to liaise with the local brigade and arrange visits to enable them to become familiar with the layout of the site and with the location of such fire critical items as main electrical switchgear, stores of highly flammable and toxic materials, dry riser installations, fire hydrants, access restrictions etc. This enables them to plan their course of action before they arrive at the site in the event of a call-out to a fire.

It is also helpful if the brigade can meet staff who have particular fire duties and who can meet them at the entrance to the premises and direct them to the site of the fire.

References

1. British Standards Institution, BS EN 3, *Portable fire extinguishers*, 9 parts, BSI, London
2. British Standards Institution, BS 7863, *Recommendations for colour coding to indicate the extinguishing media contained in portable fire extinguishers*, BSI, London
3. British Standards Institution, BS5306, *Fire extinguishing installations and equipment on premises*,
   Part 0: Guide to the selection of installed systems and other fire equipment
   Part 1: Hydrant systems, hose reels and foam inlets
   Part 2: Sprinkler systems
   Part 3: Portable fire extinguishers
   Part 4: Carbon dioxide systems
   Part 5, Section 5.1: Halon 1301 total flooding systems, BSI, London
4. Building Regulations 1991, *Approved Document No. 1*, HMSO, London
5. British Standards Institution, BS 5588, *Fire precautions in the design, construction and use of buildings* (11 parts), *Part 8, Code of practice for means of escape for disabled people*, BSI, London
6. British Standards Institution, BS 5378, *Safety signs and colours*, BSI, London
7. British Standards Institution, BS 5499, *Fire safety signs, notices and graphic symbols.*
   Part 1: Specification for fire safety signs
   Part 2: Self luminous fire safety signs
   Part 3: Internally illuminated fire safety signs, BSI, London
8. European Union, Directive No. 89/391/EEC, *on the introduction of measures to encourage improvements in the safety and health of workers at work*, EU, Luxembourg (1989)
9. European Union, Directive No. 89/654/EEC, *concerning the minimum safety and health requirements for the workplace*, EU, Luxembourg (1989)

Bibliography

Statutes and Publications available from HMSO, London:
The Fire Services Act 1947 (as amended by the Acts of 1953 and 1959)
The Fire Precautions Act 1971
The Fire Safety and Safety of Places of Sport Act 1987

The Petroleum (Consolidation) Act 1928 (as modified by the Petroleum Spirit (Motor Vehicles etc.) Regulations 1929)
The Highly Flammable Liquids and Liquefied Petroleum Gases Regulations 1972
Fire Certificates (Special Premises) Regulations 1976
Fire Precautions (Workplace) Regulations 1997
Health and Safety (Safety Signs and Signals) Regulations 1996 as amended in 1996
The Chemicals (Hazard Information and Packaging for Supply) Regulations 1994

Guidance on Statutes:
Guide to fire precautions in existing places of work that require a fire certificate: factories, offices, shops and railway premises
Code of practice for fire precautions in factories, offices, shops and railway premises not required to have a fire certificate.
Guidance on fire precautions in places of work.
Fire safety at work.

Further reading

Books
King, R.W. and Magid, J., *Industrial Hazard and Safety Handbook*, Butterworth, London (1979)
Lees, F.P., *Loss Prevention in the Process Industries*, Vols 1 & 2, Butterworth, London (1996)
Underdown, G.W., *Practical Fire Precautions*, 3rd edn, Gower Press, Teakfield Ltd, Farnborough (1989)
Wharry, D.M. and Hirst, R., *Fire Technology, Chemistry and Combustion*, Institution of Fire Engineers, Leicester (1974)
Palmer, K.N., *Dust Explosions and Fires*, Chapman & Hall, London (1973)
Field, P., *Explosability Assessment of Industrial Powders and Dusts*, HMSO, London (1983)
Dewis, M. and Stranks, J., *Fire Prevention and Regulations Handbook*, Royal Society for the Prevention of Accidents, Birmingham (1989)
Portsmouth University, *Safety Technology Unit Module ST4: Fire and Explosion*, Portsmouth (1988)
Everton, A., Holyoak, J. and Allen, D., *Fire Safety and the Law*, 2nd edn, Paramount Publishing Ltd, Borehamwood (1990)
Todd, C. *Croner's Guide to Fire Safety*, Croner Publications Ltd, Kingston upon Thames (1995)

Other publications (listed according to publisher)

Fire Offices Committee, London
Classification of Fire Hazards in Buildings
Rules for Automatic Sprinkler Systems

Fire Protection Association, London
Compendium of Fire Safety Data:

Vol. 1 Management of fire risks (including organisation of fire safety)
2 Industrial and process fire safety (including occupancy fire safety)
3 Housekeeping and general fire precautions (including nature and behaviour of fire)
4 Information sheets on hazardous materials
5 Fire protection equipment and systems
6 Buildings and fire (design guide and building products)

British Standards Institution, London
BS EN 2: Classification of fires
BS 5445: Components of automatic fire detection systems (9 parts),

Part 5: Heat sensitive detectors – point detectors containing a static element,
Part 7: Specification for point type smoke detectors using scattered light, transmitted light and ionisation,
Part 8: Specification for high temperature heat detectors
Part 9: Methods of test of sensitivity to fire
BS 5839: Fire detection and alarm systems for buildings (8 parts),
Part 2: Manual call points
Part 3: Automatic release mechanisms for certain fire protection equipment
Part 4: Control and indicating equipment
Part 5: Optical beam smoke detectors
Part 6: Design and installation of fire detection and alarm systems in dwellings
Part 8: Design installation and servicing of voice alarm systems
BS 5908: 1980 Code of Practice for fire precautions in chemical plant
BS EN 50054: Electrical apparatus for the detection and measurement of combustible gases
BS 6266: 1982 Code of Practice for fire protection for electronic data processing installations (formerly CP 95)

Health and Safety Executive publications available from HSE Books, Sudbury
Guidance Notes:
CS 4 The keeping of LPG in cylinders and similar containers
CS 6 The storage and use of LPG on construction sites
CS 8 Small scale storage and display of LPG at retail premises
CS 11 Storage and use of LPG at metered estates
CS 21 The storage and handling of organic peroxides
GS 3 Fire risk in the storage and industrial use of cellular plastics
GS 16 Gaseous fire extinguishing systems: precautions for toxic and asphyxiating hazards
GS 19 General fire precautions aboard ships being fitted out or under repair
GS 20 Fire precautions in pressurised workings
GS 40 Loading and unloading of bulk flammable liquids and gases at harbours and inland waterways
PM 25 Vehicle finishing units: fire and explosion hazards

Guidance Booklets:
HS(G) 3 Highly flammable materials on construction sites
HS(G) 4 Highly flammable liquids in the paint industry
HS(G) 5 Hot work: welding and cutting on plant containing flammable materials
HS(G) 11 Flame arresters and explosion reliefs
HS(G) 22 Electrical apparatus for use in potentially explosive atmospheres
HS(G) 34 Storage of LPG at fixed installations
HS(G) 41 Petrol filling stations: construction and operation
HS(G) 50 The storage of flammable liquids in fixed tanks (up to 10 000 m^3 total capacity)
HS(G) 51 The storage of flammable liquids in containers
HS(G) 64 Assessment of fire hazards from solid materials and the precautions required for their safe storage and use: a guide for manufacturers, suppliers, storekeepers and users
HS(G) 103 Safe handling of combustible dusts
HS(G) 113 Lift trucks in hazardous areas
HS(G) 123 Working together on firework displays: A guide to safety for firework display organisers and operators
HS(G) 124 Giving your own firework display
HS(G) 131 Energetic and spontaneously combustible substances identification and safe handling
HS(G) 136 Workplace transport safety. Guide for employers
HS(G) 139 The safe use of compressed gases in welding, cutting and allied processes
HS(G) 140 The safe use and handling of flammable liquids
HS(G) 146 Dispensing petrol: Assessing and controlling the risk of fire and explosion at sites where petrol is stored and dispensed

HS(G) 158 Flame arresters: Preventing the spread of fires and explosions in equipment that contains flammable gases and vapours
HS(G) 168 Fire safety in construction
HS(G) 176 The spraying of flammable liquids

Legal Series Booklets:
L 65 Prevention of fire and explosion and emergency response on offshore installations. Offshore Installations Regulations 1995 and Approved Code of Practice and Guidance
L 16 Design and construction of vented non-pressure road tankers used for the carriage of flammable liquids

Chapter 3

Safe use of machinery

J. R. Ridley

3.1 Introduction

Machinery and equipment have been evolved to meet a need whether for producing, changing or moving materials and components. Initially they were essentially functional with scant regard for the health and safety of those using them. But attitudes have changed and all work equipment must now be designed and built so that it does not put the user at risk of damage to health or injury. However, equipment does not work by itself – it needs someone to work it, drive it or, in the case of robots, tell it what to do. It is at this interface between equipment and operator that the risks to health and injury arise.

Many techniques have been developed to reduce these risks to a minimum and this chapter looks at some of those techniques that are available to the designer and user of modern equipment. What is not dealt with is the other vital element in the interface – the operator – and the training necessary to ensure his/her safety and how it matches the equipment, the culture and the working methods of the particular organisation. Essentially the term *work equipment* encompasses any equipment used in the course of work. However, in this chapter work equipment will be considered in four major functional areas: machinery, power trucks, cranes and lifts, and pressure systems.

3.1.1 Legislative arrangements

With the gradual demise of the FA and the influence of the EU, UK health and safety legislation concerning work equipment has polarised into that dealing with the facilities provided with 'new' equipment where the emphasis is on the provision of safeguards, and that dealing with the use of all work equipment which extends to include the provision of safeguards for 'existing' (pre-1993) equipment. In both these cases the relevant UK legislation stems from EU directives which rely on *harmonised EN standards*, or where they do not exist, extant national standards, to specify the conditions to be met for conformity with the directive.

The relevant legislation is the Machinery Directive[1] and its UK manifestation – the Supply of Machinery (Safety) Regulations 1992 (SMSR) – and the Work Equipment Directive[2] with its UK counterpart, the Provision and Use of Work Equipment Regulations 1998 (PUWER 2) respectively.

3.1.2 Machinery directive

When the first Machinery Directive was adopted, it excluded a number of special purpose machines because of pressure from specialist sectors on the grounds that they had special needs that could not be covered in the general safety requirements that applied to run-of-the-mill machines. To prevent delaying adoption of the directive these exclusions were allowed. However, since 1993 the European Commission has studied these particular cases and amended the directive to include additional conditions having specific application so that these particular cases could be brought within its compass.

The Machinery Directive was drawn up using the *new approach to legislative harmonisation* whereby the main body of the directive itself lays down broad objectives to be achieved, lists in annexes the areas to which safety attention should be directed and relies on European harmonised (EN) standards to specify the conditions to give conformity.

3.1.3 The Supply of Machinery (Safety) Regulations 1992

These Regulations, which incorporate the contents of the Machinery Directive, lay down the requirements to be met by the manufacturers of new machinery, i.e. machinery which is currently being put on the market, and indicate what a purchaser of a new machine, whether manufactured in the UK or elsewhere in the EU, can expect by way of safeguards on the machine and information about it from the supporting documentation. Their aim is to ensure that the machinery meets the standards necessary to ensure safety in use. Where machinery conforms with the requirements of the Regulations (and hence the directive) it will have open access to the whole of the EU market. Compliance with these conditions is also required of machinery imported into the EU. However, the requirements do not apply to machinery that is to be exported to non-EU countries.

The Regulations cover a wide range of machinery within the definition (reg. 4):

(a) an assembly of linked parts or components, at least one of which moves including . . . the appropriate actuators, control and power circuits, joined together for a specific application, in particular for the processing, treating, moving or packaging of a material
(b) an assembly of machines, that is to say, an assembly of items of machinery which . . . in order to achieve the same end are arranged and controlled so that they function as an integral whole . . ., or

(c) interchangeable equipment modifying the function of the machine . . .

but reg. 5 refers to a list of exclusions which includes, *inter alia*, machinery whose risks are covered by other directives plus lifting equipment for raising and/or moving persons. However, requirements for construction of lifts, both for goods and people, are now covered by the Lifts Regulations 1997.

General duties are put on suppliers of machinery to conform with these Regulations, whether the machine is manufactured in the UK or imported from a non-EU country (reg. 11). Documentary evidence, in the form of a technical file and certificates, is required to prove that the machinery conforms to the Regulations. The procedures for preparing the documentation are laid down and vary according to whether the machine is:

- constructed to EN or equivalent standards
- constructed to other equally effective safety standards or
- machines posing special hazards and which are listed in Schedule 4.

These procedures are shown diagrammatically in *Figure 3.1*.

An essential common element in the conformity assessment procedure is the preparation of the technical file which should contain:

- drawings of the machine
- a list of:
  - the relevant essential health and safety requirements
  - transposed, harmonised or other standards complied with
  - technical specifications
- a description of the safety devices incorporated
- copies of any test reports
- operating instructions.

Where a number of machines of the same type are to be made, only one technical file need be prepared but the manufacturer must provide documentary evidence of the procedure he will follow to ensure that all machines are manufactured to the same standard.

For all machines, except those posing high risks, where the machinery has been designed and manufactured to comply with EN or transposed national standards, the manufacturer completes a Certificate of Conformity and attaches the CE mark to the machine (reg. 13) (Figure 3.2).

In the case of the high risk machines that are listed in Schedule 4 and which comply with an EN or transposed standard (reg. 14), the manufacturer can either:

- send a copy of the technical file to an approved body for its retention or
- submit the technical file to an approved body requesting:
  - verification that the standards have been correctly applied and
  - the issue of a Certificate of Adequacy or

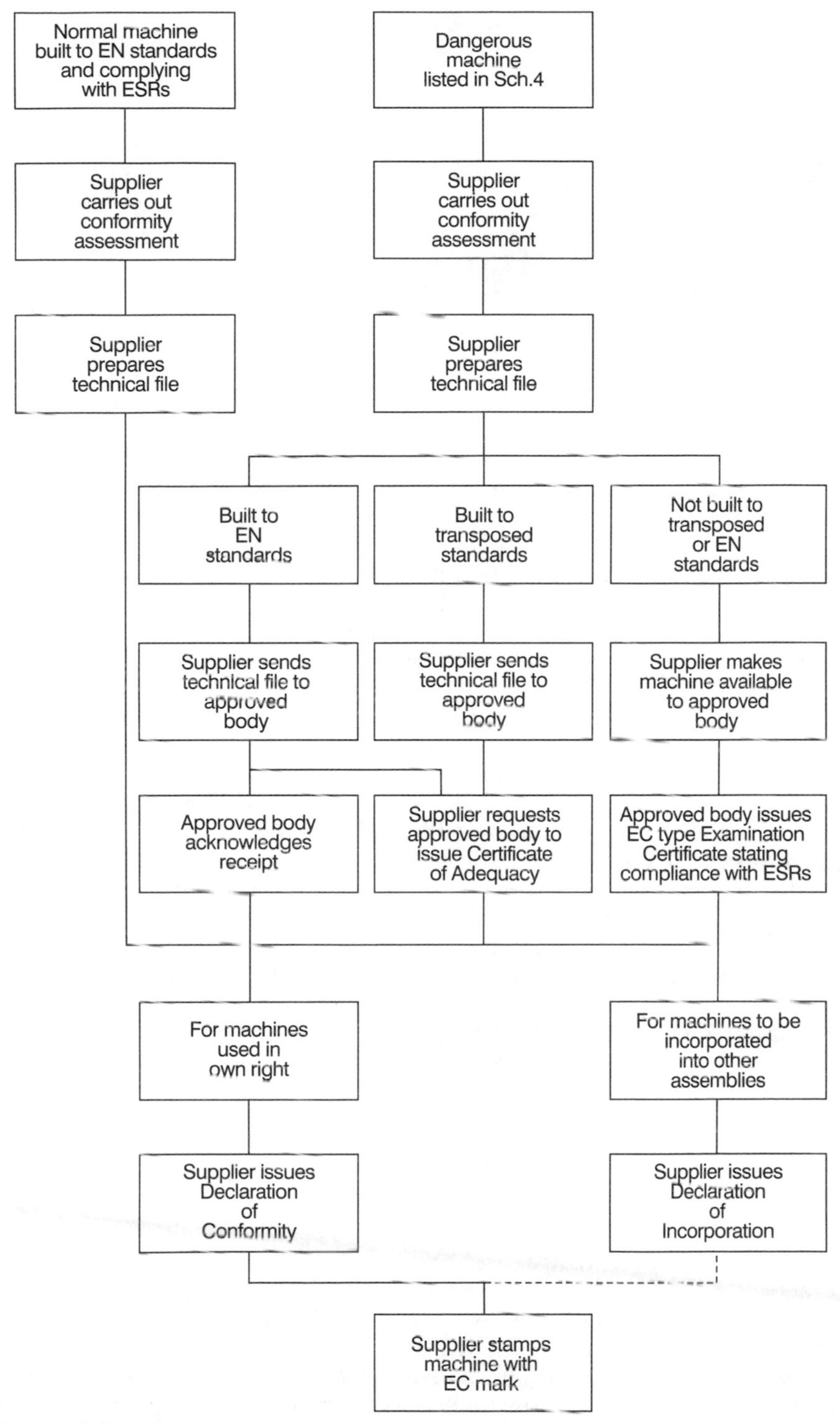
Normal machine built to EN standards and complying with ESRs
Dangerous machine listed in Sch.4
Supplier carries out conformity assessment
Supplier carries out conformity assessment
Supplier prepares technical file
Supplier prepares technical file
Built to EN standards
Built to transposed standards
Not built to transposed or EN standards
Supplier sends technical file to approved body
Supplier sends technical file to approved body
Supplier makes machine available to approved body
Approved body acknowledges receipt
Supplier requests approved body to issue Certificate of Adequacy
Approved body issues EC type Examination Certificate stating compliance with ESRs
For machines used in own right
For machines to be incorporated into other assemblies
Supplier issues Declaration of Conformity
Supplier issues Declaration of Incorporation
Supplier stamps machine with EC mark

Figure 3.1

Figure 3.2

- submit the technical file to an approved body, arrange for an example of the machine to be available for EC-type testing and request an EC Type-examination Certificate.

Where a high risk machine is not manufactured to harmonised standards or no harmonised standards exist, an example of the machine must be made available to the approved body who must satisfy themselves that the machine complies with the relevant essential health and safety requirements and can be used with safety. They will then issue an EC Type-examination Certificate. The details to be included in the various certificates are laid down in the Regulations.

An 'approved body' is appointed by a Member State (in the UK by the Secretary of State) and notified to the European Commission as an organisation which has the qualifications and necessary resources to undertake the examination and certification work. Approved bodies may charge the manufacturer fees for the certification work it undertakes on his behalf (reg. 19).

While the body of the regulation lays down the procedures to be followed for ensuring conformity with the required standards, it relies on a list of ESRs given in Schedule 3, to detail specific aspects requiring attention. Evidence of conformity with any of the ESRs is through compliance with the appropriate harmonised EN standard.

3.1.4 The Provision and Use of Work Equipment Regulations 1998 (PUWER 2)

These Regulations, which are supported by an Approved Code of Practice[3] (ACoP), are concerned with safety in the use of all work equipment and extend to lay down safeguarding requirements for existing equipment, i.e. that which had been purchased for use at work before 31 December 1992, and is not covered by SMSR. Although these Regulations have been made under HSW certain of the detail Regulations demand higher levels of protection than *so far as is reasonably practicable* and expect *foreseeable* hazards to be anticipated in other cases the test of *practicability* has to be applied, i.e. costs do not come into the consideration. However, in general the level of protection demanded is no greater than that required by earlier, but now superseded, UK legislation, i.e. FA and associated regulations. Guidance on complying with the requirements is contained in an Approved Code of Practice (ACoP)[3].

The Regulations cover work equipment (reg. 1) which the ACoP describes as anything provided for use at work, from a scalpel to scaffolding, a ruler to a reactor but excludes livestock, chemical substances, building structures and private cars. They cover all work situations except shipboard activities on a sea-going ship (reg. 3). The responsibility for compliance is placed on employers (reg. 4) and extends to the self-employed, anyone who has control of non-domestic premises used for work and to anyone who under the FA would be an occupier. Employers must ensure that all work equipment they provide is suitable for the intended work (reg. 5) and is used only for that purpose. In providing the equipment, employers must carry out an assessment to identify reasonably foreseeable risks to operators taking into account the equipment and the working conditions. If the equipment is adapted to a further purpose, the employer must ensure that it is suitable for its new use, can be used safely and that no new risks are introduced.

All work equipment must be kept well maintained (reg. 6) and in the case of machines a maintenance log must be kept. This could form the basis for a system of planned maintenance, covering routine maintenance such as lubrication and cleaning as well as examinations and overhauls, thus ensuring that machines are maintained before they fail or become dangerous. Where the operation, servicing or maintenance of any work equipment involves specific hazards (reg. 7) the employer must ensure that the work is carried out safely and that only properly trained and competent operators are allowed to carry out the particular tasks.

Suitable information and instruction must be given to those who use the work equipment but the obligation has been extended to cover foreseeable abnormal use of the equipment (reg. 8). Any such information and instruction must be comprehensible to the recipients making allowance for different assimilation abilities and also those whose first language is not English. Similarly, operators must be trained (reg. 9) to use the equipment safely and any supervision provided must be familiar with and competent in its safe use.

Where the protective measures demanded for work equipment under other relevant EU directives differ from that demanded by regs 11 to 24, the higher standard applies (reg. 10). But this still allows two identical machines, one purchased before 31 December 1992 and one purchased after that date, to be subject to different legal standards of safeguarding. Regulation 11 lays down a hierarchy of measures to be taken to protect against dangerous parts of machinery. First, it places *absolute* requirements on employers to prevent access to rotating stock bars and to provide means to ensure that dangerous parts of machinery have stopped before contact can be had with them. The hierarchy of measures demand the highest level of protection that it is *practicable* to achieve in the following priority order:

1 fixed guards to EN standards[4],
2 other guards or devices that prevent contact with dangerous parts. Many parts are only dangerous when moving, so measures, such as interlocking guards that stop the machine before contact can be made, are acceptable. EN standards[4,5] cover suitable means as does the obsolescent BS 5304[6],
3 the use of jigs, holders, push sticks etc. that keeps the operators hands away from the dangerous parts. These are particularly relevant in the use of woodworking machinery,
4 systems of work requiring the provision of information, training, instruction and supervision. In the UK, systems of work have never been accepted as primary means of protection against dangerous parts of machinery but only as back-up for other safeguards.

The *practicablity* condition does not allow consideration of cost but expects practices to match the state-of-the-art knowledge. Wherever a decision is made to use a lesser standard of protection, the reasons for selecting it should be recorded since the decision may later be called into question and have to be justified, possibly before a court.

Any guards provided shall:

(a) be suitable for their intended purpose
(b) be of good construction, sound material and adequate strength
(c) be kept in good repair and effective working order
(d) not introduce hazards
(e) not be easily defeated
(f) be an adequate distance from the dangerous parts[7,8,9]
(g) not interfere with the operation of the equipment
(h) allow maintenance to be carried out safely.

The intent of earlier law often required interpretation in the courts especially where a legal action following an accident revolved round compliance with a particular section or phrase[10]. In reg. 12 the opportunity has been taken to bring the substance of some of those decisions into statute law covering such matters as:

(a) articles or substances falling or being ejected from the machine
(b) rupture or disintegration of the work equipment[11]
(c) equipment overheating or catching fire
(d) the unintended or premature discharge from the equipment of articles or substances
(e) explosion of the equipment or anything in it.

The measures to be taken, which exclude the use of personal protective equipment and systems of work, are aimed at preventing or minimising the effects of any hazards likely to be met in the workplace. Also excluded from these requirements are processes covered by other specific legislation, such as that dealing with lead, asbestos, radiations, noise and hazardous chemicals.

Employers are required to provide protection against high and low temperatures (reg. 13), and ensure machinery has adequate controls:

- for starting or changing the state of the equipment (reg. 14)
- to stop the equipment (reg. 15)
- for emergency stopping (reg. 16)
- located so operator can see all of the machine or, if not, incorporate an audible warning with time delay start (reg. 17)
- that do not cause dangers even under fault conditions (reg. 18).

All powered work equipment should be capable of being isolated from its source of power (reg. 19). This is a requirement of the Electricity at Work Regulations 1989 but PUWER 2 extends it to include hydraulic, pneumatic and any other source of power. When used, work equipment must be stable either inherently, by clamping or by using stabilising devices such as the outriggers on mobile cranes and on access towers (reg. 20), and be sufficiently well lit to enable the work to be carried out without risk to health or safety (reg. 21).

Maintenance work should be carried out with the work equipment shut down and locked off. However, if maintenance work must be done with the machine running, all necessary precautions must be taken to protect those carrying out the work (reg. 22). Where work equipment presents a danger to health and safety it should be clearly identified as such (reg. 23) and, if appropriate, be fitted with warning devices (reg. 34).

Guidance on the techniques and practices for achieving compliance with the requirements of these Regulations can be found in the ACoP[3].

3.2 Strategy for selecting safeguards

At the design stage of new work equipment, when considering the purchase of additional equipment or the alteration of existing equipment allowance must be made for the safeguards necessary to ensure safety, or continuing safety, of the equipment and any consequent effect on the working methods to be followed to ensure its safe use. To this end, a strategy should be developed that includes the following stages:

1 Identify the hazards.
2 Eliminate the hazards or reduce to a minimum.
3 Carry out a risk assessment of the residual hazards.
4 Design/select safeguards.

5 Develop safe operating methods.
6 Inform operators of hazards and train in safe operating methods.
7 Install safeguards.
8 Monitor effectiveness and acceptability of safeguards and modify as necessary.

These stages are considered below.

3.2.1 Identifying the hazards

The foreseeable hazards at all stages of the equipment's use need to be taken into account and include:

(a) its physical dimensions
(b) method of drive and power requirements
(c) parameters of speed, pressure, temperature, size of cut, mobility etc.
(d) materials to be processed or handled and method of feed
(e) operator position and controls
(f) access for setting, adjustments and maintenance
(g) environmental factors such as dust, fumes, noise, temperature, humidity etc.
(h) operating requirements including what the operator needs to do.

Thus an overall picture can be obtained of any limitations or constraints on the design of suitable safeguards and give a pointer to providing the most effective measures for the safe operation of the equipment.

Typical hazards that can be met include:

- crushing
- shearing
- cutting or severing
- entanglement
- drawing-in or trapping
- impact
- stabbing or puncture
- friction or abrasion
- high pressure fluid ejection
- electrical shock
- noise and vibration
- contact with extremes of temperature.

Identification of these hazards can be from personal knowledge of the equipment, from the descriptions given in EN 292.1[12] or using one of the techniques described in annex B of EN 1050[14]. Reports of incidents involving other similar types of equipment can also prove a useful source of information.

3.2.2 Eliminating or reducing hazards to a minimum

Examples of the elimination or reduction of hazards include:

- changing process material for something less hazardous
- modifying the process
- reducing the operating limits of speed, pressure, temperature, power etc.
- automating the production or handling process
- controlling the process from a remote safe position.

3.2.3 Assessment of residual risks

A risk assessment is defined[12] as:

> A comprehensive estimation of the probability and the degree of possible injury or damage to health in a hazardous situation in order to select appropriate safety measures:

Evaluation of the risk requires an estimation of the likely severity of any injury or damage and of the probability of it happening. A procedure for obtaining information needed to make the assessment of the risk is outlined in EN 1050[14].

The assessment of lesser risks tends to be qualitative with the assessor making a subjective evaluation. However, for more complex risks or where there are a number of parallel risks some form of quantitative assessment may be necessary to determine priorities. Typical techniques that can be employed are described in annex B of EN 1050 and include:

- Preliminary hazard analysis (PHA)
- 'What-if' method
- Failure mode and effects analysis (FMEA)
- Fault simulation for control systems
- Method organised for a systematic analysis of risk (MOSAR)
- Fault tree analysis (FTA)
- DELPHI technique.

There is also a growing body of reliability data for component items of work equipment which can be used in a quantitative assessment of the risk.

3.2.4 Design and selection of safeguarding measures

Where the hazard cannot be eliminated it will be necessary to provide safeguarding measures which for machinery could be any of the techniques described in section 3.1.4. Selection of the type of safeguard will depend on a number of factors including:

- the operating methods and systems of work for the equipment
- proximity of the operator to hazardous areas or points
- need for access for cleaning, setting, adjustment, maintenance etc.
- stopping time of the machine
- the severity of the potential injury or ill-health from the residual hazard.

3.2.5 Operator training

Where working methods change because of the introduction of new equipment or as a result of modifications to existing equipment, the operators should be trained in any new techniques involved. Ideally the operators should be involved in the process of making changes, whether to new equipment or of the addition of guards to existing equipment, since this can facilitate its introduction. Once a decision has been taken concerning the new measures to be adopted, a programme of training and instruction can be developed and implemented so that when the changes are put into effect, the operators are familiar with the new operating methods and procedures.

3.2.6 Monitoring the effectiveness of the safeguards

Once the safeguarding measures have been installed the operations should be monitored to ensure that:

- the operators are following the new work method
- the safeguards do not interfere with the control of the process
- the safeguards provide an adequate degree of protection
- the safeguards are robust enough and are not likely to fail.

3.3 Safeguarding techniques

In new machines, guards should be designed in as an inherent part of the machine while in existing machines any added guards should be designed to provide the necessary protection while allowing the machine to be operated with the minimum of disruption. There is a wide range of types of guards and guarding techniques on which a designer can draw and these are outlined in EN standards[12,13] with design requirements specified in BS EN 953[4]. In these standards the use of the word 'guard' implies a physical barrier whereas 'safety device' means other non-physical measures for providing the desired level of protection, such as interlocks, pressure sensitive trips, electro-sensitive protective devices (photo-electric curtains) etc. Safeguard is a general term that refers to the means provided to protect against access to dangerous parts and can be either a guard or a safety device or a combination of both.

3.3.1 Fixed guards

A fixed guard is defined[12] as a guard that is kept in place permanently by welding or by fasteners that can only be released by the use of a tool and when in position it should not be capable of being displaced casually. Where fixed guards need to be removed periodically for maintenance or clearing a jam-up they could be a hinged door or a slot-in panel secured by a bolt or other suitable device. The simpler the device the more likely is it to be used especially if it is backed by commonsense, training and good supervision.

3.3.2 Distance guard

A distance guard is simply a barrier sited at an appropriate distance from the danger. The degree of risk being faced will determine whether a fixed rail or fence is necessary; in the latter case, the distance from the dangerous part will determine the opening or mesh sizes, or vice versa. Guidance on safe distances is given in Appendix A of a BS[6] and two EN standards[7,15].

3.3.3 Adjustable guards

Adjustable guards comprise a fixed guard with adjustable elements that the setter or operator has to position to suit the job being worked on. They are widely used for woodworking and toolroom machines. Where adjustable guards are used, the operators should be fully trained in how to adjust them so that full protective benefit can be obtained.

3.3.4 Tunnel guard

Where there is an automatic feed to or delivery from a dangerous part of a machine, operator safety can be provided through the use of tunnel guards which should be of a cross-section to permit the free movement of the product and long enough (at least 1 m) to prevent the operator reaching the dangerous part. If metal components are being delivered from a machine and the chute is lined with acoustic material an additional benefit will be obtained from the reduction in noise generated.

3.3.5 Fixed enclosing guard

The provision of a fixed guard enclosing the whole of a dangerous machine achieves a high standard of guarding. Work can be fed to the machine through an opening in the guard by a manual or automatic device arranged so that the operator cannot reach the dangerous parts.

3.3.6 Interlocked guards

Where guards need to be moved or opened frequently and it is inconvenient to fix them, they can be interlocked mechanically, electrically, pneumatically, etc., to the machine controls. Two basic criteria must be observed: until the guard is closed the machine should not be capable of being started; and the machine should be brought to rest as soon as the guard is opened. Where there is a run down time, the guard may need to be fitted with a delay release mechanism. The interlock system can provide either control interlocking which acts through the machine controls, power interlocking that operates by interrupting the primary power supply or by mechanical disconnection of the machine from its power source. Different arrangements of interlocking systems are reviewed in BS EN 1088[5].

An essential feature of an interlock or safety circuit is that it must be completed before the machine can start and that any break in it trips the controls and brings the machine to rest. With hydraulic and pneumatic safety circuits, the circuit must be pressurised in the safe condition, any loss of pressure causing the system to trip.

Interlocked guards can be hinged, sliding or removable but the integrity of the design of the interlocking mechanisms is crucial. The mechanisms must be reliable, capable of resisting interference and the system should fail safe.

Interlocking guards allow ready access while ensuring the safety of the operator. However, there are circumstances that may require the machine to be moved when the guard is open, i.e. for setting, cleaning, removing jams, etc. Such movement is only permitted under the following circumstances:

1 As part of a 'permit-to-work' system.
2 On true inch control.
3 On limited inch control with each movement of the producing part not exceeding 75 mm (3 in) at a predetermined minimum speed.
4 If for technical reasons the machine or process cannot accept intermittent movement, a continuous movement is permitted provided it

is only at a predetermined minimum speed and is controlled by a hold-on switch, release of which causes the machine to stop immediately. Also, there should be only one such control operable on a machine at any one time.

The choice of interlocking method will depend on power supply and drive arrangement to the machine, the degree of the risk being protected against and the consequences of failure of the safety device. The system chosen should be as direct and as simple as possible. Complex systems can be potentially unreliable, have unforeseen fail-to-danger elements and are often difficult to understand, inspect and maintain, and can have low operator acceptability.

3.3.6.1 Types of interlocks

(a) Direct manual switch or valve interlocks (*Figure 3.3*) where the switch or valve controlling the power source cannot be operated until the guard is closed, and the guard cannot be opened at any time the switch is in the run position.
(b) Mechanical interlocks provide a direct mechanical linkage from the guard to the power transmission shaft. The most common application is on power presses (*Figure 3.4*).
(c) Cam-operated limit switch interlocks are versatile, effective and difficult to defeat. They can be rotary (*Figure 3.5*) or linear (*Figure 3.6*) and in each case the critical feature is that in the safe operating position the switch is relaxed, i.e. the switch plunger is not depressed. Any movement of the guard from the safe position causes the switch plunger to be depressed, breaking the safety circuit and stopping the machine. This is the 'positive mode' of operation (*Figure 3.6(b)*) and must be used whenever there is only one interlock switch. 'Negative mode' of operation (*Figure 3.6(a)*) occurs when the switch plunger is depressed as the guard moves to the safe position and is not acceptable for single switch applications. However, a combination of the two in series is used on high risk machines such as injection moulding machines. This arrangement can incorporate a switch condition monitoring circuit as shown in *Figure 3.7*. The type of electrical switch used in interlocking is important. They must be of

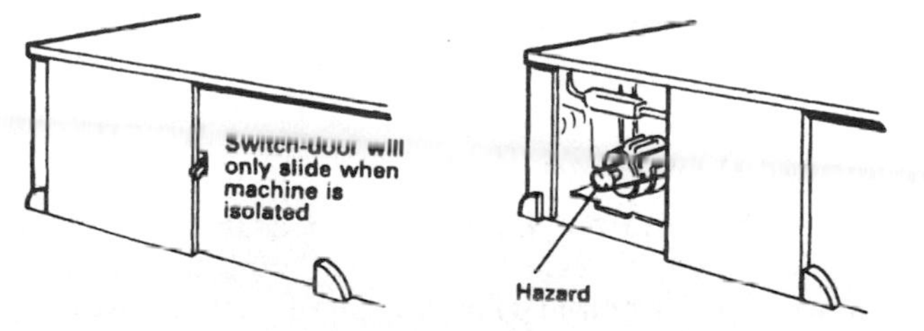

Figure 3.3 Direct manual switch interlock. (Detail design is crucial to ensure an effective arrangement.) (Courtesy Engineering)

Figure 3.4 Interlocking guard, mechanically linked to a crankshaft arrestor fitted to a power press. (Courtesy J. P. Udal Company)

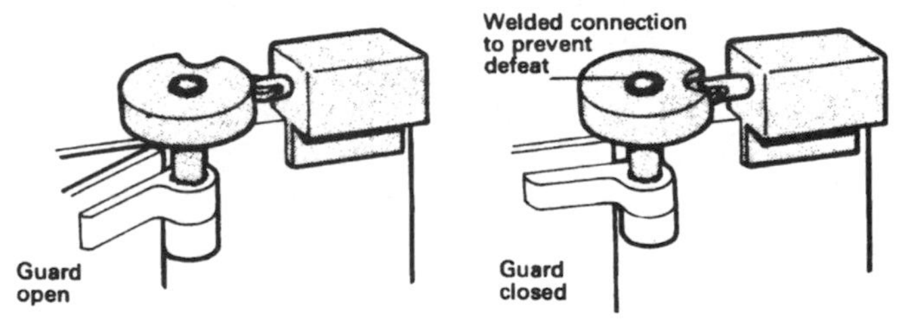

Figure 3.5 Cam activated limit switch for a hinged guard (positive) operation. (Courtesy Engineering)

the positive make-and-break type (known as 'limit' switches) that fail to safety and have contacts capable of carrying the maximum current in the circuit. The principle of operation of such switches is shown in *Figure 3.8*. Micro-switches relying on leaf spring deflection for contact breaking are not acceptable.

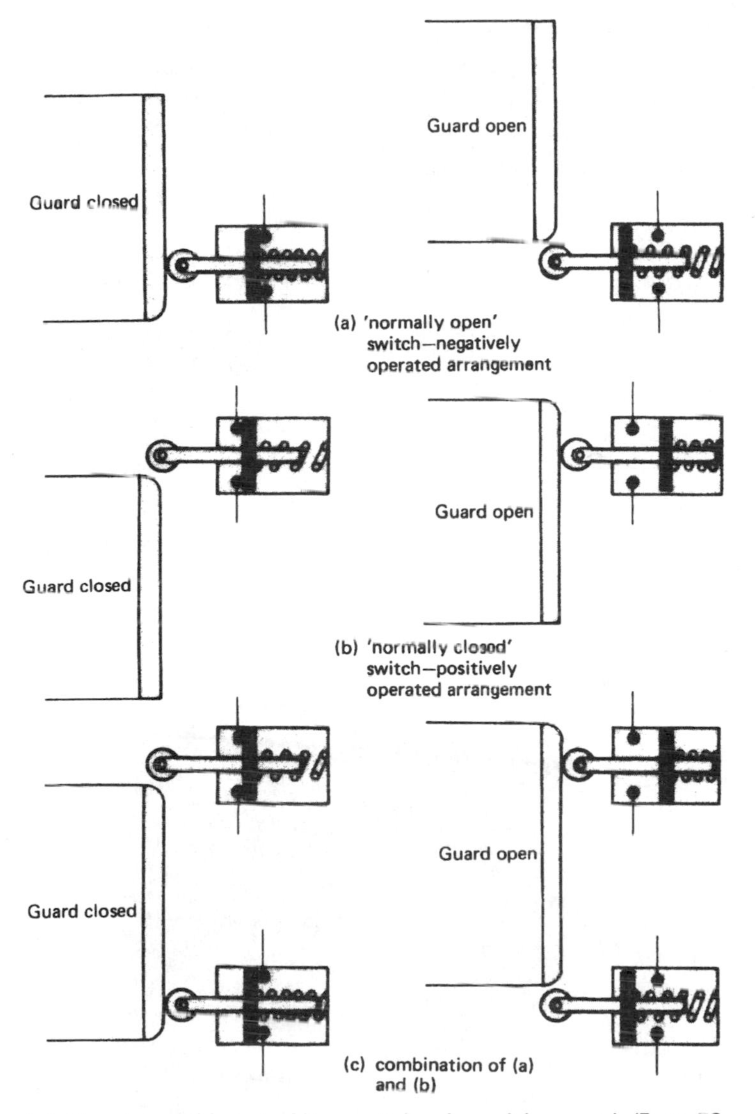

Figure 3.6 Mounting and layout of limit switches for a sliding guard. (From BS 5304:1988)

(d) Trapped key interlocks (key exchange system) work on the principle that the master key, which controls the power supply to the machine through a switch at the master key box, has to be turned OFF before the keys for individual guards can be released. The master switch cannot be turned to ON until all the individual keys are replaced in the master box. Each individual key will enable its particular guard to be opened, releasing the guard key which the operator should take with him when he enters the machine. The

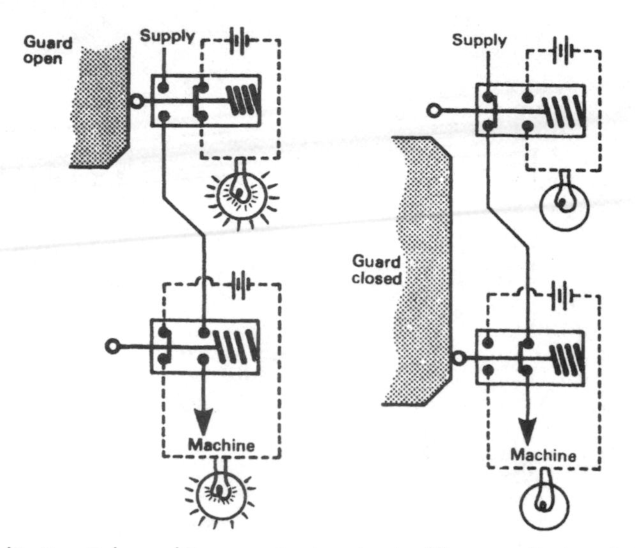

Figure 3.7 Limit switch condition monitoring circuit. (Courtesy Engineering)

Figure 3.8 Limit switch with cover removed showing the principle of fail-safe operation. (Courtesy Dewhurst & Partner plc)

individual key is trapped in the guard lock until the guard is replaced and locked by the guard key. A diagram of a typical installation is shown in *Figure 3.9*.

(e) Captive key interlocking involves a combination of an electrical switch and a mechanical lock in a single assembly where usually

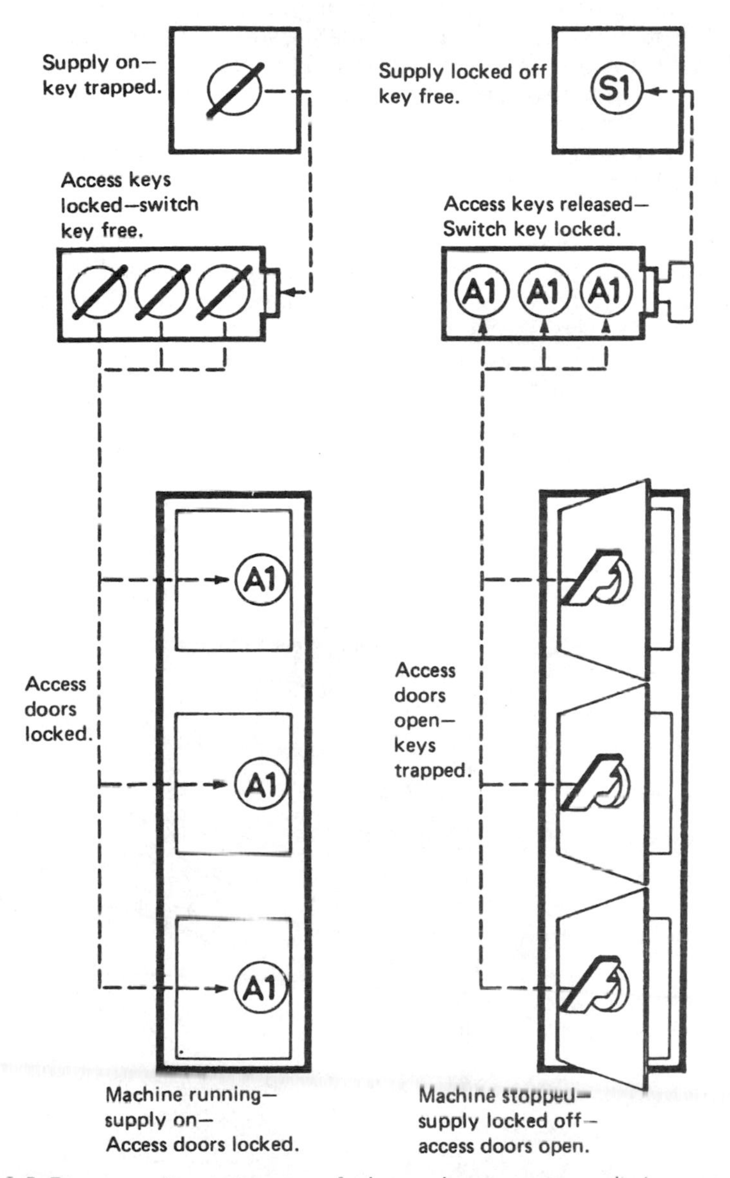

Figure 3.9 Diagrammatic arrangement of a key exchange system applied to a machine with a number of access doors required to be open at the same time. (Courtesy Castell Safety International Ltd)

the key is attached to the movable part of the guard. When the guard is closed, the key locates on the switch spindle. First movement of the key mechanically locks the guard shut and further movement actuates the electrical switch to complete the safety circuit (*Figure 3.10*).

(f) The type of magnetic switch shown in *Figure 3.11* which uses a number of magnets uniquely configured to match components of the switch part, provides a high degree of protection. It also has the advantage, since it is encapsulated, of withstanding washing, a benefit in the food industry. Other non-contact switches work through inductive circuits between an actuator and the switch.

(g) Time delay arrangements are necessary when the machine being guarded has a large inertia and, consequently, a long rundown time on stopping. An electromechanical device is shown in *Figure 3.12* where the first movement of the bolt trips the machine, but the bolt has to be unscrewed a considerable distance before the guard is released. A solenoid-operated bolt can also be used in conjunction with a time delay circuit that is actuated from either the machine controls or the trip circuit.

(h) Mechanical scotches are required on certain types of presses to protect the operator when reaching between the platens. These

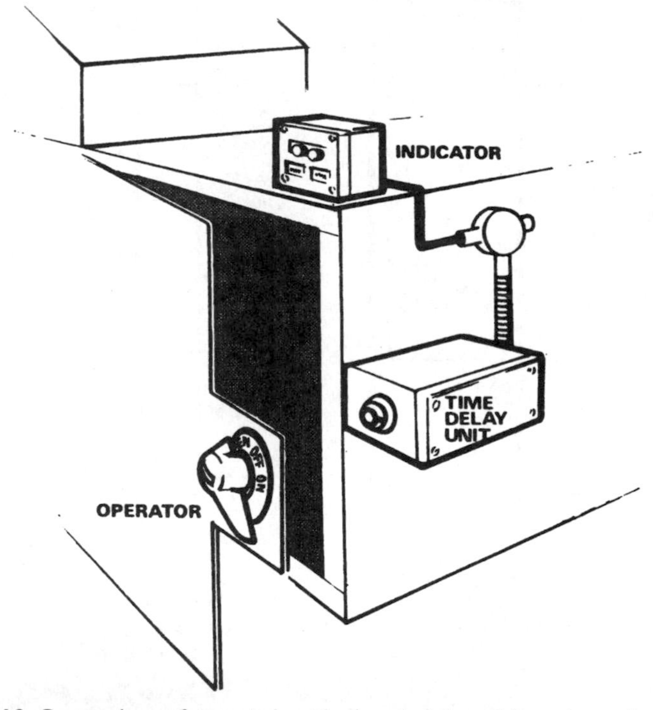

Figure 3.10 Captive key safety switch with electrical time delay release device. (Courtesy Unimax Switch Ltd)

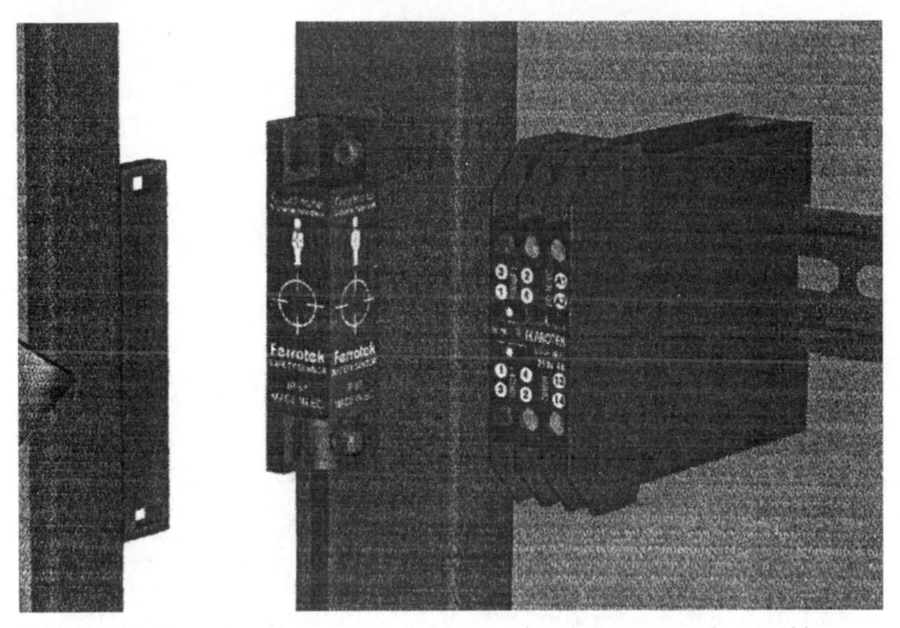

Figure 3.11 Magnetic switch with coded magnet actuator, sensor and control box (Courtesy Guardmaster Ltd)

scotches can be linked to the guard operation so that they are automatically positioned each time the guard is opened. Similar devices are needed to restrain the raised platform of a tipper lorry when work is done on the chassis. However, with scissor lifts the scotch must be inserted between the bottom rollers and the base frame.

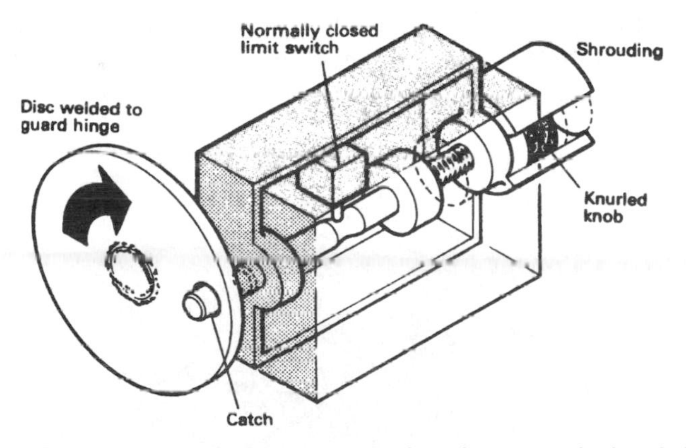

Figure 3.12 Electro-mechanical delay device; a shroud prevents the knurled knob from being spun too quickly. (Courtesy Engineering)

3.3.7 Automatic guard

This type of guard closes automatically when the machine cycle is initiated and is arranged so that the machine will not move until the guard is in the safe position. The machine then commences its movement automatically. Where trapping points occur as the parts of the guard come together, trip devices should be fitted. A version of this type of guard moves across the danger zone as the machine operation is initiated, removing any part of the operator that is in that area. This type is sometimes known as a 'sweep-away' guard.

3.3.8 Trip devices

A trip device is any device which, as the operator approaches the dangerous part, automatically trips the safety circuit. A simple flap trip is shown in *Figure 3.13* but trip devices can include: trip bars and wires, sensitive probes (on drilling machines), photoelectric devices, pressure sensitive strips and mats, and emergency stop switches. *Figure 3.14* shows a pull trip switch on a conveyor. It is important that trip devices are properly adjusted and that the machine's brake is in good working order.

The safe distance of trip barriers and electrosensitive screens from the dangerous part depend on the speed at which a person can move into the danger area and the rapidity with which the dangerous parts can be brought to rest. An EN standard[9] gives data on the approach speeds of parts of the body.

3.3.9 Two-hand control devices

Two-hand controls are used on machines having a single cycle operation where the work is placed in the machine and the machine struck on. They are applicable only to machines with a single operator. It is

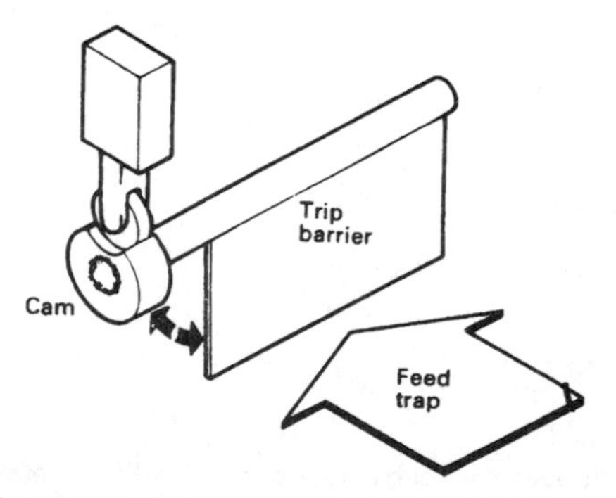

Figure 3.13 Hinged barrier with cam-operated limit switch. (Courtesy Engineering)

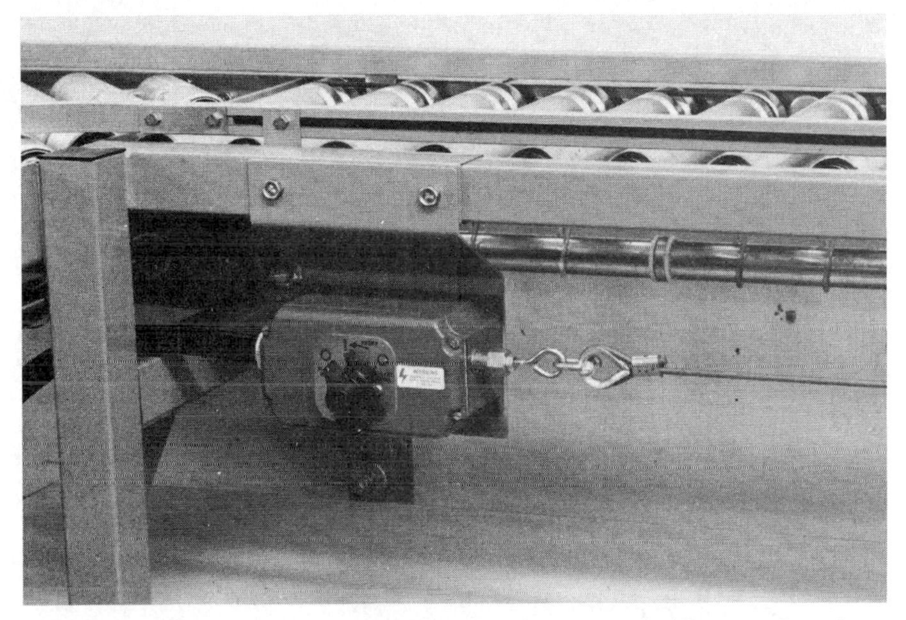

Figure 3.14 Roller conveyor with emergency grab wire switch. (Courtesy Craig & Derricott Ltd)

essential that the control buttons are positioned more than a hand-span apart, the control circuit arranged so that both controls must be activated simultaneously to start the cycle and that both controls must be released after each cycle before the next cycle can be initiated. Release of either button during the dangerous part of the cycle must stop or reverse the machine movement. Details are specified in an EN standard[16].

3.3.10 Control guards

On some small machines, such as forming presses and gold blocking machines, in which the guard has to be moved out of the way to place and remove the work, the guard can be arranged so that as it is moved to the safe position it actuates the machine cycle, and all the time the machine is moving the guard is locked in position. This type of guard also has application on some large panel presses.

3.3.11 Guard material

To ensure a high standard of protection is afforded, guard material must have sufficient strength and durability and not interfere with the functioning of the machine. Typical materials include:

- sheet steel – first choice of material especially where fluids such as lubricating oil need to be contained
- weld mesh or expanded metal – useful where ventilation is required
- polycarbonate – gives clear view of work, used extensively in the food industry since it can easily be cleaned by washing. It is extremely tough but has the disadvantage of scratching easily.

3.3.12 Openings in guard materials

Where a guard material has openings or gaps in it, care must be exercised to ensure that the guard is positioned such that it is not possible for any parts of the human body to reach the dangerous parts. The size and shape of the opening are important in determining what parts of the body can pass through. Information on the relationship between opening size and shape and distance from the danger area is contained in publication BS 5304[6] and BS EN standard[17].

3.3.13 Reaching over guards

When employing a distance guard or fence it is important that the height of the fence and its distance from the dangerous parts is sufficient to ensure the dangerous parts cannot be reached over the top of the guard. Information on the safe distance and heights is given in two BS EN standards[7,15].

3.4 Powered trucks

Powered trucks, which come within the equipment encompassed by PUWER 2, cover a vast range of equipment from small pedestrian operated units through the various sizes of counterbalance and reach fork-lift trucks, rough terrain trucks to the enormous straddle vehicles used for lifting and transporting 40 tonne containers. Each of these types has its own operating techniques but there is one common aspect applicable to them all – training of the operators. In general, these trucks are powerful and expensive. In inexperienced hands a great deal of costly damage can be done to and by the trucks.

3.4.1 Training truck operators

Training should be to a formalised programme and the trainers should be skilled in the operation of the particular type of truck, well versed in teaching skills and the training organisation should have the necessary resources. Advice on training is given in an Approved Code of Practice[18].

It is equally important that the candidates selected for training should be suitable and in particular they should:

(i) be over 18 years of age and under 60
(ii) be medically examined and passed as fit to operate the particular type of truck
(iii) have stereoscopic vision (with glasses if necessary). Monocular vision (one eye only) can cause difficulties in estimating distances
(iv) not be colour blind particularly where stacking and storage is by colour coding
(v) not be known illegal drug users.

Three major aspects of training that should be covered are:

1 Theory of operation of trucks, particularly the effects of rear wheel steering.
2 Familiarisation with the truck and its controls.
3 Operating in the working environment.

At the end of the training, the candidates should be examined on their knowledge and skills and, if found to have reached a satisfactory standard, be given a certification of competence or 'licence'. The licence should give details of the truck or trucks which the operator is competent to drive.

3.4.2 Conditions for the safe operation of powered trucks

Although rider operated reach and counterbalance trucks are the most common in use, the operating conditions that apply to them also largely apply to other types of truck. Legislative requirements for the safe operation of mobile work equipment are contained in part III of PUWER 2 and include:

1 Controls and structure of the truck should give good all-round visibility.
2 The truck should be fitted with overhead and back guards as shown in *Figure 3.15*.
3 State of the truck, its cleanliness and mechanical condition.
4 Floor surface – level and in good condition without potholes, capable of withstanding the point loading of truck wheels and well drained. Gullies should be covered by substantial well fitting grills or bridging plates.
5 Lighting must be adequate, whether outside or within buildings with particular attention being paid to the level of lighting in racking aisles. Care should be taken to avoid glare and areas of high light and shadow contrast. When operating away from lit areas, the truck should have its own lights. Guidance on lighting is given in publications by CIBSE[19] and HSE[20].

Figure 3.15 High lift fork truck with load rest, overhead and rear guards. (Courtesy Hyster Europe Ltd)

6 There must be adequate space to allow trucks to manoeuvre between stacking positions for positioning and recovery of loads and between vehicles being loaded or unloaded. *Figure 3.16* shows an empirical method for calculating the manoeuvring space a truck requires for placing loads.
7 Where ramps or slopes have to be negotiated the trucks should always move straight up or down the incline. Turning on or moving across a slope can result in the truck becoming unstable. When travelling up or down a slope, the load must be above the driver, i.e. towards the top of the slope.
8 Loads must be lowered as far as possible before being transported and never transported in the raised position.

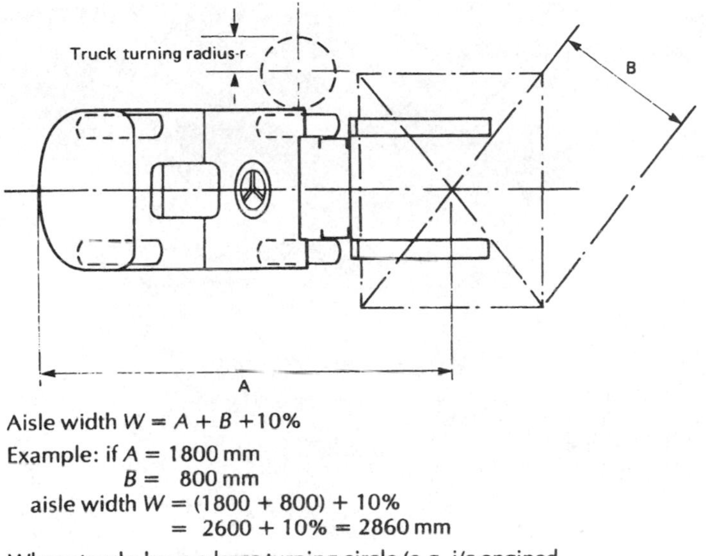

Aisle width $W = A + B + 10\%$

Example: if $A = 1800$ mm
$B = 800$ mm
aisle width $W = (1800 + 800) + 10\%$
$= 2600 + 10\% = 2860$ mm

Where trucks have a large turning circle (e.g. i/c engined trucks) the inner turning radius 'r' must be added, thus aisle width $W_1 = (A + B + r) + 10\%$

Figure 3.16 Width of stacking aisle for fork trucks

9 Pedestrians should not be allowed in areas where ride-on trucks are operating. Truck and pedestrian routes should be separated and suitably marked, as shown in *Figure 3.17*.
10 However, in areas where pedestrians and trucks cannot be separated such as stores and picking areas, priority right of way should be given to pedestrians.
11 The maker's specified maximum loads should never be exceeded.
12 Ignition keys or start cards should not be left in unattended trucks to prevent use by untrained employees. However, some companies train all employees in an area as truck operators and provide trucks as a facility to be used by all.
13 Multi-item loads should be tied or bonded to prevent load movement during travel, as shown in *Figure 3.18*.
14 Passengers must not be carried unless the truck is fitted with a purpose made seat.
15 Adequate high level ventilation should be provided in the charging areas of battery electric trucks.
16 Where powered trucks are used for lifting people, special working platforms should be used and the advice given in HSE guidance note[21] followed.

Guidance on the safe operation of lift trucks is given in an HSE Guidance Booklet[22]. The main responsibility for the safe operation of powered trucks lies with the operator.

Figure 3.17 Separation of truck and pedestrian gangways

Figure 3.18 Bonded and wrapped pallet load

3.5 Lifting equipment

The term lifting equipment covers any equipment used in the process of lifting loads or people and includes lifts, cranes, hoists and lifting accessories that join the load to the crane. The legislative requirements for the construction of new lifting equipment and for its day-to-day use are contained in two separate sets of regulations. The construction of new lifting equipment is subject to SMSR supplemented by the Lifts Regulations 1997 (LR) which impose additional requirements that are specific to lifts. In use, lifting equipment is subject to PUWER 2 supplemented by the Lifting Operations and Lifting Equipment Regulations 1998 (LOLER).

3.5.1 Definitions of lifting equipment

Cranes are any lifting machine and associated parts where the movement of the load is not restricted by guides or rails. The description includes permanent fixed installations in buildings such as overhead travelling cranes, temporary installations typically found on construction and building sites, self-contained mobile cranes and hand-operated chain pulley blocks.

Lifting accessories are defined in SMSR as:

> '. . . components or equipment not attached to the machine and placed between the machinery and the load or on the load in order to attach it;'

and *separate lifting accessories* as:

> '. . . accessories which help to make up or use a slinging device, such as eyehooks, shackles, rings, eyebolts, etc.'

Lifts are defined in LR as:

> '. . . an appliance serving specific levels, having a car moving
> (a) along guides which are rigid, and
> (b) along a fixed course even where it does not move along guides which are rigid (for example, a scissor lift),
> and inclined at an angle of more than 15 degrees to the horizontal and intended for the transport of:
> – persons
> – persons and goods
> – goods alone if the car is accessible, that is to say, a person may enter it without difficulty, and fitted with controls situated inside the car or within reach of a person inside.'

3.5.2 Construction of lifting equipment

3.5.2.1 Cranes

Cranes and their accessories are work equipment and as such must be designed and manufactured to conform with SMSR with supporting documentation as evidence of conformity. In addition, before they are put into service they must be subjected to the tests summarised in *Table 3.1*. The supplier should issue a Test Certificate on completion of the test.

Table 3.1. Table of test coefficients for lifting equipment

Lifting equipment	Test coefficients	
	Static	*Dynamic*
Powered equipment	1.25	1.1
Manual equipment	1.5	1.1
Lifting ropes	5	
Lifting chains	4	
Separate accessories:		
Metallic rope eyes	5	
Welded link chains	4	
Textile ropes and slings	7	
Metallic components of slings	4	

3.5.2.2 Lifts

The manufacture of lifts follows the same procedural requirements as other work equipment but with the added requirements contained in the Lifts Regulations (LR) which recognises the value of quality assurance schemes and also the fact that many of the components may be supplied by specialist manufacturers. Similar obligations are placed on both the lift manufacturer (reg. 8) and the component manufacturer (reg. 9) to ensure their products meet the required standard. These obligations include:

- lifting equipment and components must satisfy the appropriate ESRs. Evidence of this is through compliance with a harmonised (EN) standard
- carrying out a conformity assessment
- drawing up a Declaration of Conformity
- affixing the CE mark to the inside of the lift car or the component
- ensuring it is, in fact, safe.

The Lifts Regulations also recognise the important role quality assurance schemes play in ensuring high standard of product, and consequently safety, and use it as a core requirement in the conformity assessment procedure. The conformity assessment (reg. 13) is undertaken by 'notified bodies' who:

- may carry out unannounced inspections during manufacture
- examine and check details of the quality assurance scheme under which the lift or component was manufactured
- carry out a final inspection.

In an alternate certification procedure, the lift maker can request the notified body to carry out a 'unit verification' on his product to confirm that it conforms to the requirements of the Regulations.

Where a quality assurance scheme has been part of the manufacturing process of a lift but the design has not been to harmonised standards, the manufacturer can request the notified body to check that the design complies with the requirements of the Lifts Directive[23].

Notified bodies (reg. 16) are bodies or organisations with suitable technical and administrative resources to carry out inspections and conformity assessments. They are appointed by the Secretary of State who notifies the European Commission and their appointment is published in the Official Journal of the EU. When a lift is being installed, the builder and the installer are responsible for ensuring that the lift shaft contains no pipework or cabling other than that necessary for the operation of the lift (reg. 11).

In addition to the ESRs contained in SMSR, lifts must meet the ESRs listed in the Lifts Regulations which include:

- Take precautions to prevent the car falling, such as double suspension ropes or chains, the incorporation of an arrester device and means to support the car in the event of a power or control failure.
- Ensure the functions of the controls are clearly indicated and that they can be reached easily, especially by disabled persons.
- The doors of the car and at the landings must be interlocked to prevent movement of the car when any of the doors are open or prevent any doors being opened except when a car is at the landing.
- Access to the lift shaft must not be possible except for maintenance or in an emergency and there must be arrangements at the ends of travel to prevent crushing.
- The car must be provided with:
  - suitable lighting
  - means to enable trapped persons to be rescued
  - a two-way communication system to contact emergency services
  - adequate ventilation for the maximum allowed number of passengers
  - a notice stating the maximum number of passengers to be carried.

For information on the detailed requirements, the Regulations and appropriate EN standards should be consulted.

3.5.3 Safe use of lifting equipment

The requirements to be met for the safe use of lifting equipment are contained in PUWER 2 supplemented by the Lifting Operations and Lifting Equipment Regulations 1998 (LOLER) and a supporting

Approved Code of Practice[24]. These Regulations cover all work equipment for lifting loads including accessories that connect the load to the crane and they revoke the Hoists Exemption Order 1962. A load is defined to include persons (reg. 2). These Regulations are proscriptive and risk based and require the carrying out of risk assessments of lifting operations.

The obligations imposed on the employer (reg. 3) have been extended to the self-employed, to anyone who has control of lifting equipment and to anyone who controls the way lifting equipment is used.

Lifting equipment must be suitable for its purpose (reg. 4) and constructed of materials of adequate strength with a suitable factor of safety taking account of any hostile working environment. It should be stable when used for its intended purpose and this is particularly pertinent for mobile lifting equipment which should be provided with outriggers. Access to operating positions and, where necessary, other parts should be safe and precautions should be taken to prevent slips, trips and falls whether on the equipment itself or when moving in the work area during a lifting operation. Protection must be provided for the operator especially where he is likely to be exposed to adverse weather. Instruments should be provided to detect dangerous weather conditions such as high winds so precautions can be taken and, if necessary, the equipment taken out of use.

Additional measures have to be taken for lifts that carry people (reg. 5) including enhanced strength of lifting ropes, means to prevent crushing or trapping, falling from a carrier and to allow escape from a carrier in an emergency. The lifting equipment should be positioned to minimise the risk of equipment or load striking someone (reg. 6), loads should not be carried over people and hooks should have safety catches. Where carriers pass through shafts or openings in floors they should be fenced to prevent anyone falling through.

The safe working load or maximum number of passengers, as appropriate, should be marked on all lifting equipment (reg. 7). All lifting operations should be properly planned and supervised (reg. 8) and measures taken to ensure that no loads pass over places where people are working and that people do not work under suspended loads. The operator should have a clear view of the load or be directed by a banksman using signs or signals clearly understood by himself and the operator. Lifting equipment should not be used for operations likely to cause it to overturn, for dragging loads or used in excess of its safe working load. Lifting accessories should be used within their safe working loads and stored where they will not deteriorate or be damaged.

All lifting equipment must be regularly inspected (reg. 9) to a programme laid down either as a result of an assessment of its use or based on past experience. The inspection should be carried out by someone competent and knowledgeable in the equipment – such as an insurance surveyor – and a report containing the prescribed particulars prepared for the employer. Any faults affecting the safe operation must be reported to the enforcing authority (reg. 10). Reports of inspections and documents accompanying new equipment must be kept available for inspection (reg. 11).

The requirements of these Regulations are more flexible in implementation than earlier prescriptive requirements and allow realistic duties to be developed to match the actual conditions of use.

3.5.3.1 Safe use of cranes

Perhaps the most commonly used piece of handling equipment is the crane, which over the years has been developed to meet highly specialised applications, with the result that there is now a great range of types and sizes in use in industry, the docks and on construction sites. As a result of accidents in the past, a body of legislation has grown up which covers the construction and use of cranes. This body of legislation has been consolidated into SMSR for the design and manufacture and LOLER for the safe use and periodic inspections of cranes. There are a number of common techniques and safety devices that contribute to the safe operation of cranes and some of these are summarised below:

Overtravel switches
To prevent the hook or sheave block from being raised right up to the cable drum, a robust limit switch should be fitted to the crab or upper sheave block. Checks of this limit switch should be included in routine inspections.

Protection of bare conductors
Where bare pick-up conductors are used to carry the power supply they must be shielded from accidental contact particularly if near cabin access. Suitably worded notices, e.g. WARNING – BARE LIVE WIRES, should be posted on the walls or building structure. The power supply isolating switch should be provided with means for locking-off during maintenance work.

Controls
The controls of cranes, whether cabin, pendant or radio, should be clearly identified to prevent inadvertent operation. On overhead electric travelling (OET) cranes with electric pendant controls the directions of travel should be unambiguously marked. Controls should be of the 'dead-man' type.

Load indicators
Load indicators are required to be fitted to jib cranes and can be used with benefit on all cranes.

Safety catches
Crane hooks should be fitted with safety catches to prevent slings, chains, ropes etc. from 'jumping' off the hook.

Emergency escape
Where, on travelling cranes, access to the cab is not an integral part of the crane, suitable escape equipment should be provided to enable the driver to reach the ground quickly and safely in an emergency.

Access
Safe means of access should be provided to enable:

1 the driver to reach his operating position;
2 the necessary inspections and maintenance work to be carried out safely.

Operating position
The arrangement of the driver's cab should ensure:

1 a clear view of the operating area and loads;
2 all controls are easily reached by the driver without the need for excessive movement of arms or legs;
3 all controls are clearly marked as to their function and method of operation.

Passengers
No one, other than the driver, should be allowed on the crane when it is operating unless there is a special reason for being there and it has been authorised. 'Riding the hook' is prohibited but should it be necessary to carry persons, the properly designed and approved chair or cradle should be used.

Safe working load
All cranes should be marked with their safe working load which must never be exceeded except for test purposes. If there is any doubt of the weight to be lifted, advice should be sought.

Controlling crane lifts
With many cranes including overhead electric travelling, mobile jib and construction tower cranes, the safe moving of loads relies on team effort involving the driver, slinger and sometimes a separate signaller (or banksman). Only one person, the signaller or if there is no signaller the slinger, should give signals to the driver and these should be clearly understood by both. The basic signals[25] are shown in *Figure 3.19*.

Slingers, signallers and drivers should be properly trained, medically fit and of a steady disposition. Detailed advice on the safe use of cranes, lifting accessories and mobile cranes is given by Dickie, Short and Hudson[26,27,28].

3.6 Pressure systems and gas containers

Pressure systems refer to any system of pipes, vessels, valves or other equipment for containing or transferring gases and liquid at high pressure. The requirements are contained in the Pressure Systems and Transportable Gas Containers Regulations 1989 which consider static installations and transportable containers separately. The Regulations define (reg. 2) a pressure system as:

Figure 3.19 Crane signals. (BS 7121)

(a) a system comprising one or more pressure vessels of rigid construction, any associated pipework and protective devices;
(b) the pipework with its protective devices to which a transportable container is, or is intended to be, connected; or
(c) a pipeline and its protective devices;

which contains or is liable to contain a relevant fluid, but does not include a transportable gas container.

A transportable gas container is defined as:

> a container, including any permanent fitting of such a container, which is used, or is intended to be used, to contain a relevant fluid and is –
>
> (a) designed to be transportable for the purpose of refilling and has an internal volume of at least 0.5 litres and not greater than 3000 litres;
> (b) a non-refillable container having an internal volume of at least 1.4 litres and not greater than 5.0 litres; or
> (c) for the purposes of regulation 17(3) only, a non-refillable container.
>
> A relevant fluid is defined as:
>
> (a) steam
> (b) any fluid or mixture of fluids which is at a pressure greater than 0.5 bar above atmospheric pressure, and which fluid, or mixture of fluids, is –
> (i) a gas, or
> (ii) a liquid which would have a vapour pressure greater than 0.5 bar above atmospheric when in equilibrium with its vapour at either the actual temperature of the liquid or 17.5° Celsius; or
> (c) a gas dissolved under pressure in a solvent contained in a porous substance at ambient temperature and which could be released from the solvent without the application of heat.

For both pressure systems and transportable gas containers, the designer, manufacturer and supplier are required to ensure that equipment is properly designed, constructed from suitable material and capable of being examined without danger (reg. 4). The equipment must be complete with protective devices and where means for pressure release are provided they must operate without causing danger. Any modifications to the equipment must not cause the equipment to become dangerous.

3.6.1 Pressure systems

The supplier of a pressure system must provide sufficient information concerning the design and operating limits to enable the system to be operated and maintained safely (reg. 5). All vessels must be marked, either on the vessel itself or by an attached plate, with the information listed in Schedule 4 of the Regulations, which includes an identifying number and operating parameters. The owners and users of pressure systems must ensure that they are not operated outside the safe operating limits (reg. 7). Pressure systems must not be run until a written scheme of examinations has been prepared by a competent person (reg. 8). The scheme should be risk based taking into account the operating conditions, likely periods of service and other conditions that could affect the safe operation of the system. The owner or user must ensure that the scheme is followed (reg. 9). Any examinations should be carried out by a competent person, normally this will be a specialist from the company that insures the system, who should provide a report of the examination. Where repairs or modifications are required to ensure the continuing safe operation of the system, the report should contain details of what requires to be done and give a time limit. If the equipment is found to give rise to imminent danger unless certain repairs are carried out (reg. 10), the owner or user must be informed immediately and a copy of the report sent to the enforcing authority within 14 days. In such case, the owner or user must ensure the system is not operated until the necessary repairs have been made. Adequate operating instructions should be provided to the system operators (reg. 11) and the system must be kept well maintained (reg. 12). Copies must be kept of the documentation provided by the suppliers and of the reports of examinations (reg. 13) and these must be passed on to subsequent owners and users if the equipment is disposed of. Steam receivers intended to run at atmospheric pressure must not be allowed to become pressurised (reg. 14).

3.6.2 Transportable gas containers

Very strict controls exist for the design, manufacture, modification and filling of gas containers. The standards to which a container has been designed must be verified by an approved body, either by supporting documents or by stamping on the container (reg. 16). When containers are filled (reg. 17), checks should be made to ensure that they are:

1 safe to be refilled
2 being filled with the correct gas
3 filled within the safe operating limits
4 not over filled or any excess fluid is removed safely.

The owner of a container must ensure it is examined by a competent person at appropriate intervals (reg. 18) but no details are given of who specifies the scheme of examination. An employer must not allow any employee to modify a gas container (reg. 19) unless that person is competent to do so (reg. 20) and following such modification the container must not be used until it is certified by a competent person as being safe for use. Also, no one other than a competent person may re-rate the capacity of a gas container (reg. 21). Records of the design specification of gas containers must be kept by the manufacturer and the hirer of leased equipment (reg. 22).

In any prosecution, defences are allowed that (reg. 23):

1 The offence was caused by another person.
2 All due diligence was exercised to avoid the offence.

For either of these defences to be effective, supportive information must be submitted to the prosecutor at least seven days before the hearing.

Experiences in the past have shown the injury and damage potential of pressure vessel failures. The requirements of these regulations are based on proven effective measures to prevent those failures and are realistic in recognising the economic value of basing schemes of examinations on assessments of the risks rather than a prescriptive scheme covering all usages regardless of local conditions.

3.7 Conclusion

The mechanical handling of materials plays an essential role in industry with the benefits from it being enormous; however any laxity in its use can prove very costly in terms of damage to plant and product, and, worse, of injury to employees. Whether the equipment is mobile, as with cranes and powered trucks, or is fixed in position as conveyors are, agreed safe systems of work and procedures for using it must be followed to reduce any risk to a minimum and to ensure the best utilisation possible.

References

1. European Union, *Council Directive on the approximation of the laws of Member States relating to machinery,* No. 89/392/EEC as amended by Directive No 91/368/EEC, EU, Luxembourg (1991) and consolidated in Directive No. 98/37/EC
2. European Union, *Council Directive concerning the minimum health and safety requirements for the use of work equipment by workers at work,* as amended by Directive No. 95/63/EC, EU, Luxembourg (1995)
3. Health and Safety Executive, Legal Series booklet No. L22, *Work equipment. Provision and Use of Work Equipment Regulations 1992. Guidance of the Regulations,* HSE Books, Sudbury (1992)
4. British Standards Institution, BS EN 953, *Safety of machinery – Guards – General requirements for the design and construction of fixed and movable guards,* BSI, London (1998)
5. British Standards Institution, BS EN 1088, *Safety of machinery – Interlocking devices associated with guards – Principles for design and selection,* BSI, London (1995)
6. British Standards Institution, Obsolescent standard BS 5304, *Safety of machinery,* BSI, London (1988)
7. British Standards Institution, BS EN 294, *Safety of machinery – Safety distances to prevent danger zones being reached by upper limbs,* BSI, London (1992)
8. British Stanards Institution, BS EN 349, *Safety of machinery – Minimum gaps to avoid crushing of parts of the human body,* BSI, London (1993)
9. British Standards Institution, Proposed BS EN 999, *Safety of machinery – The positioning of protective equipment in respect of approach speeds of parts of the human body,* BSI, London (to be published)
10. Uddin *v.* Associated Portland Cement Manufacturers Ltd [1965] 2 All ER 213
11. Close *v.* Steel Company of Wales [1962] AC 367; [1961] 2 All ER 953, HL

12. British Standards Institution, BS EN 292–1, *Safety of Machinery – Basic concepts and general principles for design – Part 2: Basic terminology and methodology*, BSI, London (1991)
13. British Standards Institution, BS EN 292–2, *Safety of machinery – Basic concepts and general principles for design – Part 2: Technical principles and specifications*, BSI, London (1991) and Part 2/Al (1995)
14. British Standards Institution, BS EN 1050, *Safety of machinery – Principles for risk assessment*, BSI, London (1997)
15. British Standards Institution, BS EN 811, *Safety of machinery – Safety distances to prevent danger zones being reached by lower limbs*, BSI, London (1997)
16. British Standards Institution, BS EN 574, *Safety of machinery – Two-hand control devices – Functional aspects – Principles for design*, BSI, London (1996)
17. British Standards Institution, BS EN 547, *Safety of machinery – Human body dimensions (7 parts)*, BSI, London (parts 1 & 2, 1996)
18. Health and Safety Executive, Approved Code of Practice No. 26, *Rider operated lift trucks: Operator training*, HSE Books, Sudbury (1988)
19. Chartered Institution of Building Services Engineers, *Code of interior lighting*, CIBSE, London (1994)
20. Health and Safety Executive, Guidance booklet No. HSG 38, *Lighting at work*, HSE Books, Sudbury (1997)
21. Health and Safety Executive, Guidance Note No. PM28, *Working platforms on fork lift trucks*, HSE Books, Sudbury (1981)
22. Health and Safety Executive, Guidance booklet No. HSG 6, *Safety in working with lift trucks*, HSE Books, Sudbury (1993)
23. European Union, *Council Directive on the approximation of the laws of Member States relating to lifts*, No. 95/16/EC, EU, Luxembourg (1995)
24. Health and Safety Executive, Approved Code of Practice on *Lifting Operations and Lifting Equipment Regulations 1998*, HSE Books, Sudbury (to be published)
25. British Standards Institution, BS 7121, *Code of practice for the safe use of cranes, Part 1 – General; Part 2 – Inspection, testing and examination*, BSI, London
26. Dickie, D.E., *Lifting Tackle Manual* (Ed. Douglas Short), Butterworths, London (1981)
27. Dickie, D.E., *Crane Handbook*; (Ed. Douglas Short), Butterworth, London (1981)
28. Dickie, D.E., *Mobile Crane Manual* (Ed. Hudson, R.W.), Butterworth, London (1985)

Further reading

King, R. W. and Magid, J., *Industrial Hazard and Safety Handbook*, pp. 567–603, Butterworth-Heinemann, Oxford (1979)

Health and Safety Executive, the following publications which are available from HSE Books, Sudbury:
Regulation series booklets:
HSR 30, *Guide to the Pressure Systems and Transportable Pressure Gas. Containers Regulations 1989* (1990)
Guidance booklets:
HSG 39 *Compressed air safety* (1990)
HSG 93 *The assessment of pressure vessels operating at low temperatures* (1993)
HSG 113 *Lift trucks in potentially flammable atmospheres* (1996)
HSG 136 *Workplace transport safety: Guidance for employers* (1995)
Approved Codes of Practice:
COP 37 *Safety of pressure systems. The Pressure Systems and Transportable Gas Containers Regulations 1989: Approved Code of Practice* (1990)
COP 38 *Safety of transportable containers. The Pressure Systems and Transportable Gas Containers Regulations 1989: Approved Code of Practice* (1990)
Guidance Notes: General Series
GS9 *Road transport in factories and similar workplaces* (1992)
GS39 *Training of crane drivers and slingers* (1986)
Guidance Notes: Plant and Machinery Series
PM7 *Lifts: thorough examination and testing* (1982)
PM8 *Passenger carrying paternosters* (1977)

PM24 *Safety in rack and pinion hoists* (1981)
PM26 *Safety at lift landings* (1981)
PM34 *Safety in the use of escalators* (1983)
PM42 *Excavators used as cranes* (1984)
PM45 *Escalators: periodic thorough examination* (1984)
PM54 *Lifting gear standards* (1985)
PM55 *Safe working with overhead travelling cranes* (1985)
PM63 *Inclined hoists used in building and construction work* (1987)

Chapter 4

Electricity

E. G. Hooper and revised by Chris Buck

4.1 Alternating and direct currents

4.1.1 Alternating current

An alternating current (ac) is induced in a conductor rotating in a magnetic field. The value of the current and its direction of flow in the conductor depends upon the relative position of the conductor to the magnetic flux. During one revolution of the conductor the induced current will increase from zero to maximum value (positive), back to zero, then to maximum value in the opposite direction (negative) and, finally, back to zero again having completed one cycle. A graph plotted to show the variation of this current with time follows a standard sine wave. The number of cycles completed per second, each comprising one positive and one negative half cycle, is referred to as the frequency of the supply, measured in hertz (Hz). Mains electricity is supplied in the UK as ac at a nominal frequency of 50 Hz (50 cycles per second).

4.1.2 Direct current

Direct current (dc) has a constant positive value above zero and flows in one direction only, unlike ac. A simple example of direct current is that produced by a standard dry battery.

DC is really ac which has its positive or negative surges rectified to provide the uni-directional flow. In a dc generator the natural ac produced is rectified to dc by the commutator.

DC will be found in industry in the form of battery supplies for electrically powered works plant, such as fork lift trucks, together with associated battery charging equipment. Otherwise, dc, obtained by rectification of the mains ac supply, will be encountered only for specialist applications, e.g. electroplating.

Danger from electricity may arise irrespective of whether it is ac or dc. Where dc is derived from ac supply the process of rectification will result in some superimposed ripple from the original ac waveform. Where this exceeds 10%, the electrical shock hazard must be considered to be the same as for an ac supply of equivalent voltage. Additionally, both ac and dc can cause injury as a result of short circuit flashover. The dangers of electricity are discussed in more detail later in this chapter.

4.2 Electricity supply

Alternating current electricity is generated in thermal power stations (coal-, gas- and oil-fired) and also at nuclear power stations under the control of the privatised generating and distributing companies. It is then transmitted by way of overhead lines at 400 kV (i.e. at 400 000 volts) (*Figure 4.1*), 275 kV or 132 kV to distribution substations where it is transformed down to 33 kV or 11 kV for distribution to large factories, or further transformed down to 230 V for use in domestic and commercial premises and smaller factories.

Following privatisation, many changes have and are still taking place in the electricity industry. The demand for electricity varies considerably from day to day as well as throughout each day. Generation must be matched to meet this continually varying demand, ensuring that there is always sufficient generating capacity available at minimum cost. This is achieved through arrangements operated by the electricity pool. There is now an open market for large power users (large factories etc.) to shop around suppliers to achieve the best price. The intention is that this facility will be extended soon to include all consumers.

Electricity is received by most industrial and commercial consumers as a 'three phase four wire' supply at a nominal voltage of 400/230 V. The three phases are distinguished by the standard colours red, yellow and blue, the fourth wire of the supply serving as a common neutral conductor. The individual phase voltages (230 V) are equally displaced timewise (phase displacement of 120°) and because of this the voltage across phases is higher. Three phase supplies can be used to supply both single phase equipment (with the loads balanced as equally as possible between the three phases) or three phase equipment such as to large motors.

Consumers with large energy requirements often find it more economical and convenient to receive electricity at 11 kV and to transform it down to the lower values as required. In all cases, however, it is essential that the consumer's electrical installation and equipment, both fixed and portable, meet good standards of design, construction and protection, are adequately maintained and correctly used.

The British Standards Institution[1] (BSI) issues standards and codes giving guidance on electrical safety matters. One such standard is BS 7671, otherwise known as the IEE Wiring Regulations[2]. This specifies requirements for low voltage electrical installations, i.e. those operating at voltage up to 1000 V. However, it is important to appreciate the legal obligations relating to the safe use of electricity and electrical machinery at places of work.

The Health and Safety at Work etc. Act 1974 is an enabling Act providing a legal framework for the promotion of health and safety at all places of work. Although the Act says nothing specific about electricity it does

Figures 4.1 400 kV suspension towers on the National Grid's Sizewell-Sundon 400 kV transmission line. (Courtesy National Grid)

require, among other things, the provision of safety systems and methods of work; safe means of access, egress and safe places of employment; and adequate instruction and supervision. These requirements have wide application but they are, in general terms, also relevant to the safe use of electricity. But for specific advice on the electrical legal requirements we must turn to a set of Regulations[3] made under the HSW Act.

4.3 Statutory requirements

4.3.1 The Electricity at Work Regulations 1989

The Electricity at Work Regulations 1989 (EAW) is the primary piece of legislation dealing specifically with electricity and came into force on 1 April 1990. Since these Regulations were made under the umbrella of HSW they apply in all cases where the parent Act applies. They are thus work activity, rather than premises, related and are therefore of wide application. Some of the individual regulations are relevant to all industries while others apply only to mines. Separate, but virtually identical, Regulations have been made for Northern Ireland – the Electricity at Work Regulations (Northern Ireland) 1991. The requirements of the Regulations now also apply to offshore installations by virtue of the Offshore (Electricity and Noise) Regulations 1997.

In line with modern health and safety legislation, the EAW are 'goal setting' aimed at specifying, albeit in general terms, the fundamental requirements for achieving electrical safety. Thus they provide flexibility to accommodate future electrical developments. They specify the ends to be achieved rather than the means for achieving them. With regard to the latter, guidance is provided in a number of booklets published by the HSE as well as in BSI and other authoritative guidance. The main supporting documents are:

1 a Memorandum of Guidance[4], and
2 two Approved Codes of Practice[5,6] dealing respectively with the use of electricity in mines and in quarries.

The Memorandum of Guidance referred to above gives technical and legal guidance on the Regulations and provides a source of practical help. Similarly, the two Codes of Practice provide essential advice for mines and quarries.

The Electricity at Work Regulations comprise 33 individual regulations which place firm responsibilities on employers, the self-employed, managers of mines and of quarries and employees to comply as far as they relate to matters within their control. Additionally, employees have a duty to co-operate with their employers so far as is necessary for the employers to comply with the Regulations.

Topics covered by particular regulations include:

(a) Construction and maintenance of systems, work activities and protective equipment (Reg. 4).

(b) Strength and capability of electrical equipment (Reg. 5).
(c) Adverse and hazardous environments (Reg. 6).
(d) Insulation, protection and placing of conductors (Reg. 7).
(e) Earthing and other suitable precautions (Reg. 8).
(f) Integrity of 'referenced' conductors (Reg. 9).
(g) Connections (Reg. 10).
(h) Means for protecting from excess current, for cutting off supply and isolation (Regs. 11 and 12).
(i) Precautions for work on dead equipment and on or near live conductors (Regs. 13 and 14).
(j) Working space, lighting and access (Reg. 15).
(k) Persons to be competent to prevent danger and injury (Reg. 16).
(l) Regulations applicable to mines only (Regs 17–28).

4.3.2 Status of regulations

Certain of the individual regulations are subject to the qualification 'so far as is reasonably practicable'. This means that any action contemplated should be based on a judgement balancing the perceived risk against the cost of eliminating it, or at least reducing it to an acceptable level; in other words a risk assessment.

The remaining regulations are of an 'absolute' nature, which means that their requirements must be met regardless of cost. Nevertheless, in the event of a criminal prosecution for an alleged breach of statutory duty under one of these regulations, regulation 29 allows a defence to be pleaded that all reasonable steps were taken and all due diligence was exercised to avoid the commission of the offence.

4.4 Voltage levels

Unlike their predecessors, the 1989 Regulations apply equally to all systems and equipment irrespective of the voltage level. The duty is to avoid danger and prevent injury from electricity. Voltage is but one factor determining the presence of danger and, therefore, the risk of injury; examples of other matters requiring consideration when evaluating the electrical risk are the equipment type, its standard of construction and the nature of the work environment.

4.5 Electrical accidents

Electricity is a safe and efficient form of energy and its benefits to mankind as a convenient source of lighting, heating and power are obvious. But, if electricity is misused, it can be dangerous – a statement of the obvious, but one which must be made so as to keep the matter in proper perspective.

In the UK, every year, up to 25 people may be killed, at work, as a result of an electrical accident. In addition, up to a thousand or so are injured.

These figures, considering the widespread use of electricity in industry and when compared with the numbers killed and injured as a result of other types of accident, are relatively small. Nevertheless, a knowledge of electrical safety is important because, by comparison with the proportion of serious injury resulting from accidents arising from all causes, an electrical accident is more likely to lead to serious injury. There is the potential also for expensive damage to plant and property due to fires of electrical origin, e.g the overloading of cables.

4.6. The basic electrical circuit

For an electrical current to do its job of providing lighting, heating and power, it must move safely from its source, through the conducting path and back from whence it came. In short, electric current requires a suitable circuit to assist its flow without danger. The circuit must be of suitable conducting material, e.g. copper, covered with a suitable insulating material (to stop the current 'leaking' out) such as PVC or rubber.

For an electrical current (measured in amperes) to flow in a circuit it requires pressure (measured in voltage). As it flows it encounters resistance from the circuit and apparatus and this characteristic is measured in ohms.

This relationship between volts, amps and ohms is brought together in the famous Ohm's law known to most schoolboys. Thus, to put it simply, the current in a circuit is proportional to the voltage driving it and inversely proportional to the resistance it has to overcome:

$$\text{amps} = \frac{\text{volts}}{\text{ohms}} \tag{1}$$

Alternatively, this may be written as:

$$\text{ohms} = \frac{\text{volts}}{\text{amps}} \tag{2}$$

or

$$\text{volts} = \text{amps} \times \text{ohms} \tag{3}$$

A further useful relationship is that between power (measured in watts) and the voltage and current. Thus:

$$\text{watts} = \text{volts} \times \text{amps} \tag{4}$$

From Ohm's law, this may be expressed also as:

$$\text{watts} = \text{amps}^2 \times \text{ohms} \tag{5}$$

$$\text{or watts} = \frac{\text{volts}^2}{\text{ohms}} \tag{6}$$

4.6.1 Impedance

As an alternating current passes around a circuit under the action of an applied voltage it is impeded in its flow. This may be due to the presence in the circuit of resistance, inductance or capacitance, the combined effect of which is called the impedance and is measured in ohms.

In a pure resistance circuit the applied voltage has to overcome the ohmic value of the resistance as is the case for direct current (see equations (1) to (6) above).

If, however, the circuit contains inductance, such as the presence of a coil, the alternating magnetic field set up by the alternating current will induce a voltage in the coil which will oppose the applied voltage and cause the current to lag vectorially behind the voltage (up to 90° where the circuit contains pure inductance only). This property is called reactance and is measured in ohms. Sometimes a circuit may contain a capacitor: the applied voltage 'charges' the capacitor and the effect is such as to cause the current vectorially to lead the voltage. This property is also called reactance. Now most circuits contain resistance, inductance and capacitance in various quantities, and the effect of impedance is found as follows:

$$\text{impedance}^2 = \text{resistance}^2 + \text{reactance}^2$$

Strictly speaking this is a vectorial calculation. It is beyond the scope of this section to go further into these relationships and readers are referred to a standard textbook on electricity[7] and to BS 4727[8].

4.7 Dangers from electricity

It has been said that properly used electricity is not dangerous but out of control it can cause harm, if it passes through a human body, by producing electric shock and/or burns. Electricity's heating effect can also cause fire but we will first deal with the electric shock phenomenon.

4.7.1 Electric shock

If a person is in contact with earthed metalwork or is inadequately insulated from earth then, because the human body and the earth itself are good conductors of electricity, they can form part of a circuit (albeit an abnormal one) through which electricity, under fault conditions, can flow. How can such fault conditions occur?

If, for any reason, there was a breakdown of insulation in a part of an electric circuit or in any apparatus such as, say, a hand-held metal-cased electric drill, it is conceivable that current would flow external to this supply circuit, if a path were available. For example, the metalwork of the drill may be in contact with a live internal conductor at the point of insulation breakdown (an example of indirect contact). Or, take the example of someone working at a switch or socket outlet from which the cover had been removed before the electricity supply had been isolated. In such cases the person concerned could touch live metal or a live terminal and, if the conditions were right, would thereby cause an electric current to flow through his body to earth (an example of direct contact). If the total resistance of the earth fault path were of a sufficiently low value, the current could kill or maim.

Electric shock is a term that relates to the consequences of current flow through the body's nerves, muscles and organs and thereby causing disturbance to normal function. Owing to a current's heating effect the body tissue could also be damaged by burns. A particular danger with electric shock from alternating current is that it so often causes the person concerned to maintain an involuntary grip on the live metal or conductor (particularly hand-held electric tools) and this prolongs current flow. Death could occur when the rhythm of the heart is disturbed such as to affect blood flow and hence the supply of oxygen to the brain, a condition that is known as ventricular fibrillation. Unless skilled attention is given immediately, by, say, the proper application of oral resuscitation, ventricular fibrillation can be irreversible. However, it is still fortunately the case that most electric shock victims recover without permanent disability or lasting effect.

Although the effect of a direct current shock is generally not as dangerous as with ac (there is no dangerous involuntary grip phenomenon for example), it is recommended that similar precautions against shock be taken. In any case it will be recalled from section 4.1.2 that the dc electrical shock hazard could be similar to that of an equivalent ac voltage as a result of the amount of superimposed ripple.

The severity of an electrical shock depends on a number of factors, the most significant of which are the combination of the magnitude and the duration of the flow of shock current through the body.

Personal sensitivity to electric shock varies somewhat with age, sex, heart condition etc., but for an average person the relationship between shock current, and time for which the body can accommodate it, is given by a formula of the following kind:

$$\text{current} = \frac{116}{\sqrt{\text{time}}}$$

where the current is measured in milliamps (mA) and the time is measured in seconds (s). Above the duration of one heart beat a lower current threshold is recommended.

Thus a 50 mA shock current (i.e. 0.05 A) could probably flow through a body, without much danger, for up to 4 seconds; whereas a 500 mA current (0.5 A) flowing for only 50 ms (0.05 s) could be fatal.

The maximum safe 'let go' current is less than 10 mA, whereas 20 mA to 40 mA directly across the chest could arrest respiration or restrict breathing; currents above 500 mA flowing for as little as 50 ms can be fatal. However, even 'safe' currents at the level of about 5 mA to 10 mA could still cause a minor shock sensation and cause someone to fall if working at a height.

From all this it will be concluded that at normal mains voltage of 230 V, and given the average value of resistance of a human body at 1500 Ω, the current flowing through the body would, from equation (1), be a maximum of

$$\frac{230}{1500} = 0.15\,\text{A approximately (150 mA)}$$

– a dangerously high value. Under normal circumstances there will be additional resistances (or impedances as they are more correctly called) such as, for example, the resistance of the circuit, the earth electrode, and any footwear worn. It must also be remembered that body resistance varies from person to person depending upon biological, environmental and climatic conditions. But even so, given the very small value of current that could cause harm, all possible sources of contact with live electric parts must be avoided. The precautions to be taken are discussed in later sections of this chapter.

4.7.2 Burns

Burn injuries may be associated with shock and can be seen as burn marks on the body at the points of current entry and exit or may also occur in the burning of internal tissue. However, severe burn injuries are more likely to arise as a consequence of short circuit flashover. In fact the number of fatalities arising from this latter cause is similar to that resulting directly from electric shock.

Short circuit flashovers caused during the course of live work are likely to result in serious injury for the simple reason that the worker is in close proximity to and probably directly facing the equipment that has been inadvertently short circuited. The extent of the flashover will depend on the amount of electrical energy available to flow into the fault. This will be determined by the fault level (the amount of current that the incoming electrical supply is capable of feeding into the fault) and the speed of operation of the electrical protection, e.g. a fuse or circuit breaker, to interrupt the flow of fault current.

In the case of a factory installation the fault level is likely to be of the order of several thousand amperes. This large current will be capable of generating severe arcing during the short period required for the electrical protective devices to see the fault and safely disconnect the supply. Many will be aware of the flashover capability from even low voltage dc supplies, such as a 12 V car battery, if the terminals are accidentally shorted by dropping a metal tool across them.

4.7.3 Fires

Fires may occur due to a variety of electrical problems, in particular as a result of the overheating of cables or equipment, arcing due to loose connections or the use of unsuitable electrical equipment in a flammable atmosphere. Such problems often arise due to deficiencies in the design or construction of the electrical installation or incorrect equipment specification. EAW addresses all these issues by specifying fundamental requirements to ensure that the design and construction of installations is such as to prevent, so far as is reasonably practicable, danger.

4.8 Protective means

4.8.1 Earthing and other suitable precautions

To prevent danger where it is 'reasonably foreseeable' that a conductor (other than a circuit conductor) may become charged with electricity, earthing or other suitable precautions need to be taken. In the case of earthing, all metalwork forming part of the electrical installation (metal conduit and trunking housing cables) or apparatus (metal equipment casings of switchgear, transformers, motors etc.) should be adequately and solidly connected to earth. Such earthing is provided by means of 'protective conductors' which may comprise a separate conductor, as in the 'twin and earth' cable or, where appropriate, the cable armouring or metal conduit or trunking. However, flexible or pliable conduit is not acceptable for this purpose.

It is important to ensure that the resistance of the earth return path, comprising the protective conductor and connection with earth, is as low as possible. This is to ensure that, in the event of an earth fault, there will be sufficient current to 'blow' the fuse or operate any other form of device protecting the circuit in question. The IEE Wiring Regulations (BS7671)[2] specify maximum permitted disconnection times for different types of installation. There is also a BS Code of Practice on the subject of earthing[9].

4.8.2 Work precautions

When work is to be carried out on a part of a circuit or piece of electrical equipment, certain precautions need to be taken to protect the worker concerned from electrical danger. The electricity supply should first of all be switched out, locked off and warning notices posted. This ensures that the circuit or apparatus being worked on is effectively electrically isolated and cannot become live.

Using a suitable voltage proving device, that part of the circuit to be worked on should be checked to ensure that it is dead before work is allowed to commence. Correct operation of the proving device should be confirmed immediately before and after use.

In some circumstances further precautions will need to be taken, such as earthing, to counter the effects of any stored or induced electrical charge. A permit to work system (PTW), explained in more detail in section 4.10, may also be used. Although EAW regulation 14 permits live working this must first be properly justified and then suitable precautions must be taken to prevent injury (an absolute duty!). Thus dead working is the norm and the preferred choice.

The HSE have published a number of guidance documents concerning those work activities where previous accident history has shown a need for more understanding to ensure electrical safety[10,11,12].

4.8.3 Insulation

Mention has already been made of the need to ensure that electrical conductors etc. are adequately insulated. Insulating material has extremely high resistance values to prevent electric current flowing through it. The principle of insulation is used when work has necessarily to be carried out at or near uninsulated live parts. Such parts should always be made dead if at all possible. If this cannot be done then properly trained people, competent to do the work, can make use of protective equipment (insulated tools, gloves, mats and screening materials) to prevent electrical shock and short circuit flashover. The provision and use of such equipment must meet the requirements of the Personal Protective Equipment at Work Regulations 1992 as well as regulation 4(4) of EAW. It is important that all protective equipment provided is suitable for the intended use (i.e. designed and constructed to an appropriate specification such as a British Standard), adequately maintained and properly used. A number of BSs cover the specification of such equipment[13,14,15,16].

4.8.4 Fuses

A fuse is essentially a thin wire, placed in a circuit, of such size as would melt at a predetermined value of current flow and therefore cut off the current to that circuit. Obviously a *properly rated fuse* is a most useful precaution because, in the event of abnormal conditions such as a fault, when excess current flows, the fuse would 'blow' and protect the circuit or apparatus from further damage. A fuse needs to be capable of responding to the following types of abnormal circuit conditions:

- overload
- short circuit (phase to neutral or phase to phase)
- earth fault (phase to earth).

To operate effectively and safely the fuse should be placed in the phase (live) conductor and never in the neutral conductor, otherwise even with the fuse blown or removed, parts of the circuit, such as switches or terminals, could still be live. Fuses come in various sizes with different construction characteristics and degrees of protection. Good practice advises that every fuse must be so constructed, guarded and placed as to prevent danger from such things as overheating and the scattering of hot metal when it blows. Modern cartridge fuses, the simplest variant of which is contained in the standard 13 amp fused plug, are well constructed to meet these requirements but it is difficult to tell at a glance if they have 'blown'. Simple battery continuity tests are available for easy checking and for the larger industrial sizes of cartridge fuse an automatic indication is provided.

Overfusing, that is to use a fuse rating higher than that of the circuit it is meant to protect, is dangerous because in the event of a fault a current may flow to earth without blowing the fuse. This could endanger workpeople and the circuit or apparatus concerned. In addition it could result in the cable carrying an excessive current leading to considerable overheating with the risk of fire.

4.8.5 Circuit breakers

A circuit breaker, although more expensive than a fuse, has several advantages for excess current circuit protection. The principle of operation is that excess current flow is detected electromagnetically and the mechanism of the breaker automatically trips and cuts off electricity supply to the circuit it protects. A 'blown' fuse must be replaced with one with the correct current rating whereas a circuit breaker simply needs to be reset once the fault condition has been cleared. Miniature circuit breakers (mcbs) are designed for fitting into distribution boards in place of fuses. Circuit breakers are also available to detect earth leakage current and, indeed, units are available that detect both over-current and earth leakage currents and thereby give very good circuit protection.

The majority of electric shock injuries occur when the body acts as conductor between line and earth. A general level of protection against such shocks is provided by the inclusion of a current sensitive earth leakage circuit breaker in the supply line. A typical example is shown in *Figure 4.2*.

4.8.6 Work near underground cables and overhead lines

Work near underground electricity cables and overhead electric lines has caused many serious and fatal accidents over the years. The precautions to be taken are dealt with in two HSE guidance notes[10,11].

4.9 Competency

Regulation 16 of EAW requires that no person shall be engaged in any work activity where technical knowledge or experience is necessary to prevent electrical danger or injury unless that person has the appropriate knowledge or experience having regard to the nature of the work. The Memorandum of Guidance[4] lists five factors to be considered when evaluation the scope of 'technical knowledge or experience'. These are:

- adequate knowledge of electricity
- adequate experience of electrical work

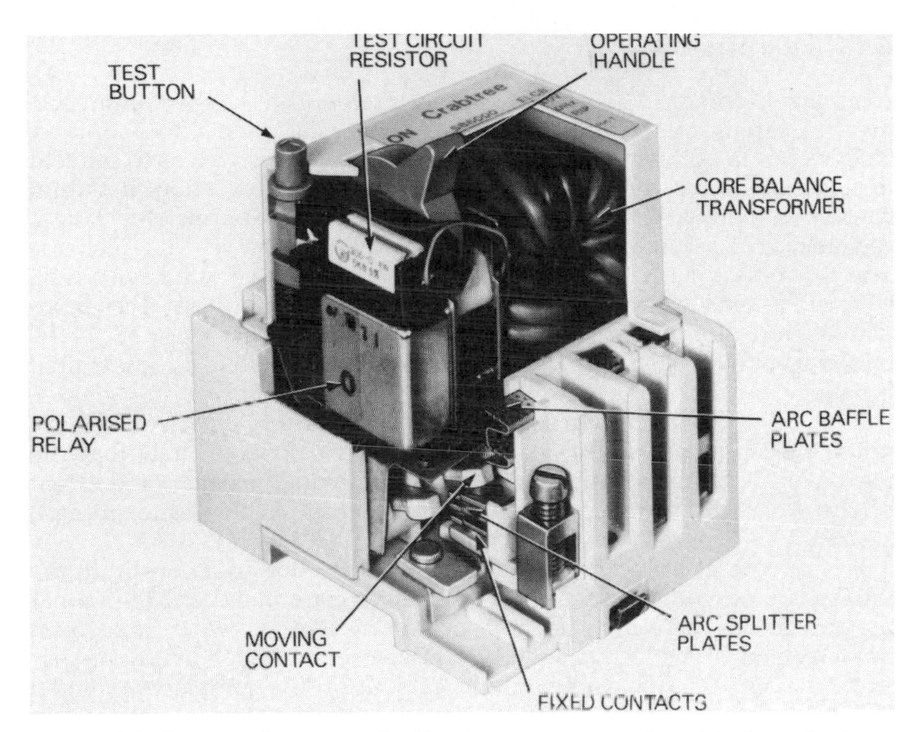

Figure 4.2 Cut-away illustration of a 30 mA current operated earth leakage circuit breaker. (Courtesy Crabtree Electrical Industries Ltd)

- adequate understanding of the equipment to be worked on
- understanding of the hazards that may arise during the work
- ability to recognise whether it is safe for work to continue.

Where technical knowledge or experience may be lacking then regulation 16 requires that the person concerned shall be under an appropriate level of supervision. The legal duty allows flexibility in that competence is required in relation to the task to be performed and the need to prevent danger and/or injury from electricity. Thus competence is not expected, nor would it be realistic to expect it, across the complete spectrum of work – only in relation to the activities in which the individual will be involved.

The five factors listed provide a framework for developing training specifications to achieve competence although it should be recognised that it is important to verify competence through assessment and monitoring on an ongoing basis. It should be noted that competence is required not only in respect of electrical work activities but also to deal with any situation where electrical danger may arise, e.g. work in the vicinity of exposed live overhead conductors or excavation close to live buried cables.

4.10 Permits-to-work

A permit-to-work (PTW) is an essential prerequisite to the commencement of certain classes of work involving special danger to people. A PTW serves to hold apparatus out of normal service as well as to prevent misunderstandings through a lack of or poor communication. It should confirm in writing what precautions have been taken (points of isolation, earthing, application of safety locks etc.) and the apparatus on which it is safe to work. A PTW is invariably used for work associated with high voltage systems (above 1000 V) and may also be helpful in other cases where there are multiple points of isolation or the work is to be undertaken by personnel other than those responsible for the initial isolation.

It is the duty of the person issuing the PTW to ensure that the necessary safety precautions detailed in it have been carried out, and that the person receiving the permit is fully conversant with the nature and extent of the work to be done. Proper arrangements for the issue, receipt, clearance and cancellation of PTWs are essential.

It is important to recognise that a PTW alone does not constitute the safe system of work but serves as a further precaution by confirming that the safeguards necessary as part of the system of work have been implemented. The effectiveness of a PTW system is dependent upon and is only as good as the safety culture existing in the company. A model PTW is shown on pages 655 and 656.

4.11 Static electricity

When two dissimilar bodies or substances meet, electrons pass from one to the other at the surface contact area. When the bodies separate, particularly if they are of an insulating material, a difference of potential occurs across the separating medium which manifests itself as static electrical charge. Such an effect can, for example, occur when there is dispersion of liquid from a nozzle, powder from a tray or paper from a reel. In the movement of flammable liquid or a powder or dust with fine particles, static electricity can be generated that can give rise to sparking of sufficient energy to ignite the vapour or dust. The precautions to be taken to prevent fire and explosion as a result of static electricity depend upon the nature of the materials and process[17]. Examples of the approach to the more common problems are as follows:

1 All pipework and containers used for conveying flammable liquids should be effectively bonded together and earthed so that any static electricity produced is immediately discharged to earth before it builds up to a dangerous energy level.
2 Workshop atmospheres where flammable solvents are used for spreading as in, say, material proofing processing, can be artificially humidified. Where practicable, specialised radioactive static eliminators or earthed metal combs near the charged material etc. can also be used. It is also a wise precaution to ensure adequate ventilation such

MODEL FORM OF PERMIT-TO-WORK – FRONT

NAME OF FIRM

PERMIT-TO-WORK

1. ISSUE NO.........................

To ....

I hereby declare that it is safe to work on the following Apparatus, which is dead, isolated from all live conductors and is connected to earth:-

ALL OTHER APPARATUS IS DANGEROUS

Points at which system is isolated ....

....

....

Caution Notices posted at ....

....

....

The apparatus is efficiently connected to earth at the following points

....

....

Other precautions ....

....

....

The following work is to be carried out ....

....

....

Signed ....
being an Authorised Person.

Time................................ Date................................

MODEL FORM OF PERMIT-TO-WORK – BACK

2. RECEIPT

I hereby declare that I accept responsibility for carrying out the work on the apparatus detailed on this Permit-to-Work and that no attempt will be made by me, or by the men under my control, to carry out work on any other apparatus.

Signed ....

Time .... Date ....

Note: After signature for the work to proceed this Receipt must be signed by and the Permit-to-Work be retained by the person in charge of the work until the work is suspended or completed and the Clearance section has been signed.

---

3. CLEARANCE

I hereby declare that the work for which this Permit-to-Work was issued is now *suspended/completed, and that all men under my charge have been withdrawn and warned that it is no longer safe to work on the apparatus specified on this Permit-to-Work, and that gear, tools and additional earthing connections are all clear.

Signed ....

Time .... Date ....

*Delete word not applicable.

4. CANCELLATION

This Permit-to-Work is hereby cancelled.

Signed ....

being an Authorised Person possessing authority to cancel a Permit-to-Work.

as to keep the gas/air mixture well below the lower explosive limit (LEL) of the flammable solvent concerned.

3 High speed rotating flat belts and pulleys are known to produce dangerous static charges and should not be used at or near flammable solvents. Where there are difficulties, special conducting belts, belt dressings and earthing of drive and pulley shafting can help to eliminate the build-up of static electricity.

4 Certain processes, such as electrostatic paint spraying, make use of the characteristics of static electricity and special precautions against solvent ignition are required[18].

4.12 Use of electricity in adverse or hazardous environments

The safe use of electricity can pose particular problems in adverse or hazardous work environments. Such environments may degrade equipment to the extent that it becomes unsafe or even develops a fault thus increasing the electric shock risk. Flammable and explosive atmospheres present special risks.

The importance of ensuring that the specification and selection of electrical equipment is appropriate to the environment and conditions of use is recognised in EAW. Regulation 6 specifically mentions various kinds of adverse or hazardous environments, namely the effects of the weather, natural hazards, temperature, pressure, wet, dusty and corrosive conditions as well as exposure to flammable or explosive dusts, vapours or gases. The requirement is that electrical equipment which may 'reasonably foreseeably' be exposed to such conditions shall be of such construction, or as necessary protected, as to prevent, so far as is reasonably practicable, danger arising from such exposure.

Work on construction sites provides a good example of circumstances where an adverse environment may be encountered. The temporary nature of, and frequent changes to, such electrical installation may also encourage wiring and equipment to be installed in an unsuitable manner without adequate protection. Because of the special risks associated with the use of electricity on construction sites the HSE has issued guidance[10,11,19]. A number of BSs are concerned with the specification of electrical equipment for use in adverse environments. BS EN 60529[20], known as the 'IP' code, provides a specification for degrees of protection provided by enclosures, such as for electrical switchgear, against the ingress of solid objects, dusts and water. BS 7375[21] provides a code of practice for the distribution of electricity on construction and building sites.

4.13 Electrical equipment in flammable atmospheres

4.13.1 Explosive and flammable atmospheres

The techniques to be adopted to prevent danger when using electrical equipment in the vicinity of potentially explosive or flammable atmospheres have changed over the years and present requirements are contained in regulation 6 of EAW.

4.13.2 Construction of equipment for use in flammable or explosive atmospheres

The construction of electrical equipment to be used where a flammable or explosive atmosphere is likely to occur must be such as to prevent ignition of that atmosphere. The selection and installation of such equipment are detailed and specialised matters requiring expert knowledge.

The relevant standards have been affected by the standard harmonisation process within the EU through the European Committee for Electrotechnical Standardisation (CENELEC) and through the International Electrotechnical Commission (IEC). BS 5345: part 2[22] covers the specification of hazardous areas while BS EN 60079–14[23] deals with electrical installations in hazardous areas other than mines. BS 5501[24] also covers this subject.

The first step in the selection of appropriate equipment will be the classification of the hazardous area from the viewpoint of the likelihood of a flammable atmosphere being present. Other considerations will include temperature class or ignition temperature of the gas or vapour involved and external influences. A number of different safeguards may be employed in the design and construction of the equipment to minimise the risk of ignition. Equipment is normally certified for use in a particular situation and marked accordingly.

4.13.3 Classification of hazardous areas

In industry, with the exception of mining, areas that are hazardous, so far as flammable gases and vapours are concerned, are classified according to the probability of occurrence of explosive concentrations of gas or vapour. These classifications, called zones, are as follows:

Zone 0 is a zone in which a flammable atmosphere is continuously present or for long periods.

Zone 1 is a zone in which a flammable atmosphere is likely to occur in normal working.

Zone 2 is a zone in which a flammable atmosphere is unlikely to occur except under abnormal conditions and then only for a short time.

The particular zone determines the types of protection required for electrical equipment in use in that zone. Ideally, the prime method of protection should be to exclude electrical apparatus from any hazardous area. However, where this is not practical or economic, the next consideration should be whether the electrical apparatus can be segregated by fire-resistant impermeable barriers. Where the installation of electrical apparatus in hazardous areas is unavoidable, the following types of protection may be used according to the circumstances:

4.13.4 Type 'n' equipment

Type 'n' equipment is so constructed that in normal operation it is not capable of igniting a surrounding explosive atmosphere. Such equipment is designed for use in Zone 2 areas.

4.13.5 Type 'e' equipment

Type 'e' equipment, also known as 'increased safety' equipment, employs a protection method to electrical apparatus that does not, in normal operation, produce sparks, arcing or excessive temperatures. Examples are transformers and squirrel cage motors. Type 'e' equipment may be used in Zone 1 areas.

4.13.6 Pressurising and purging

Pressurising is a method used whereby clean air, drawn from outside, is blown at a pressure slightly above atmospheric into the room or enclosure to maintain the atmosphere at a pressure sufficiently high to prevent ingress of the surrounding potentially flammable atmosphere.
Purging is a method whereby a flow of air or inert gas of sufficient quantity is maintained in a room or enclosure to reduce or prevent the flammable atmosphere occurring.

4.13.7 Electrical fittings

Electrical installations used in flammable atmospheres require special consideration as regards design, construction and installation of the metallic cable sheathing, cable armouring or conduit, junction boxes, cable gland sealing, wiring etc. Particular attention should be given to the earthing arrangements. This is a specialised subject for which expert opinion should be sought.

4.13.8 Intrinsically safe systems

In a circuit where the amount of electrical energy available to cause a spark is below that necessary for igniting flammable vapour or gas, the equipment is considered to be intrinsically safe. It is particularly suitable for use in instrumentation, remote control etc.

4.13.9 Flameproof equipment

Flameproof equipment is regarded as safe for use when exposed to the risk of explosive atmosphere for which certification has been given. Electrical apparatus, defined as flameproof, has an enclosure that will withstand an internal 'explosion' of the flammable vapour or gas in question which may enter the enclosure. The joints of the enclosure which are designed with clearance gaps to prevent a build-up of internal pressure also prevent any internal 'explosion' igniting vapour or gas surrounding the equipment. The surface temperature of the enclosure must be below the ignition temperature of the vapour or gas in question. Flameproof enclosures are primarily intended for use in Zone 1 or Zone 2 classification but *not* in Zone 0.

4.14 Portable tools

Portable electric tools are a convenient aid to many occupational activities. However, the necessity to use flexible cables to supply electricity to the tool introduces hazards. For example, such cables are often misused and abused resulting in damaged insulation and broken or exposed conductors. The tool itself could also become unsafe if, say, its metalwork became charged with electricity due to a fault. Constant care and adequate maintenance and storage are essential to safe use.

Of the hand-held power operated electrical tools, the most common are the electric drill and portable grinding wheel. The major safety requirements for portable tools are:

1 The supply cable should be of the flexible type (i.e. with stranded conductors) with its connections correctly made, being electrically and mechanically sound at both tool and point of supply. Plugs and sockets should be of a type appropriate to the work environment and conform to the relevant BS. The cable sheath should be clamped securely at plug and tool entries.
2 Metal cased (class 1) tools should be efficiently earthed, normally achieved by a connection from the case to the protective (earth) conductor in the supply cable.
3 For preference, tools should be powered from a low voltage supply, i.e. below mains voltage. A common arrangement in the case of a single phase supply is to provide this via a portable 230/110 V CTE step down transformer, where the output winding has a centre tap connection to earth (CTE). While the supply to the tool is 110 V any shock voltage to earth will be restricted to 55 V because of the centre point to earth connection. In the case of 110 V three phase supplies the equivalent shock voltage will be 64 V approximately.
4 Suitable means should be provided for cutting off the supply as well as for isolation purposes. In the case of portable tools, the plug and socket connection will achieve this; otherwise an isolating switch or switch-fuse will need to be installed in a readily operable position.

It is important to ensure that all portable electrical equipment is regularly inspected and adequately maintained to minimise the risk of danger to the user. Equipment of the double-insulated or all insulated types, to the relevant BS[25], has no provision for earthing and is not earthed.

4.15 Residual current devices

Additional back-up protection can be provided by residual current devices that ensure that in the event of an earth fault the current is cut off before a fatal shock is received. This form of protection works on the principle of monitoring any differential between (i) the current entering a circuit to supply power to the portable apparatus and (ii) the current returning to the supply point. For normal safe operation this current differential is zero but if there is a fault, such as leakage to earth, a differential current occurs which the device rapidly senses, tripping to cut off the supply to the apparatus.

Thus an RCD will not prevent electric shock because shock current must flow through the body to cause sufficient 'out-of-balance' between the conductors to be detected by the device. However, their sensitivity and speed of operation will limit the current flow to a few fractions of a second, thus making the shock more survivable. A typical current level that will result in operation of the device (tripping current) is 30 mA, with operation taking place within 40 ms (0.04 s). The design and construction parameters for RCDs are specified in BSs[26,27].

4.16 Maintenance

It is a requirement of EAW that all electrical apparatus and conductors shall, among other things, be adequately maintained. Maintenance does not simply mean general care and cleanliness but implies a system for regular inspection and, where appropriate, testing to ensure serviceability. For maintenance to be cost effective, its extent and frequency should be determined on the basis of an assessment of the risks, i.e. taking account of factors such as the age and type of construction of the equipment and the nature and environment of its use. Maintenance will require a thorough examination to check for signs of damage or defects. This will need to be undertaken by someone with the necessary competence, who knows what to look for, is able to recognise damage or defects that may be significant and who knows the appropriate actions that need to be taken. Electrical testing may be necessary to confirm aspects concerning the condition of the equipment that might otherwise not be possible to verify by visual examination alone.

In the case of portable equipment, testing is often carried out using a proprietary portable appliance tester (PAT). One useful test is the measurement of the insulation resistance to confirm that it is sufficiently high to prevent undue leakage. Additionally, for class 1 equipment (that must be earthed) it is important to verify that the connection to earth is sound, i.e. its electrical resistance is low, and that the conductor is capable of carrying the sort of high current that may occur under fault conditions. Both the HSE and the IEE have produced guidance on electrical equipment maintenance[28,29].

4.17 Conclusion

EAW covers everything in the life of an electrical system, from initial concept in terms of its specification and design, through the construction phase to eventual commissioning for use, its use over many years allowing for possible alteration of or addition to it during its life, to eventual dismantling at the end of its useful life. However, the legal duties are expressed only in goal setting terms or as objectives to be achieved. The means for achieving these objectives are many and varied and it is for dutyholders under the Regulations, who in general are the employer, manager of a mine or quarry, the self-employed and the individual employee, to determine and put in place measures to ensure satisfactory standards of electrical safety.

Dutyholders' responsibilities are covered in regulation 3 and require the dutyholder to take appropriate action, but only in respect of matters within his or her control. In practice, employees will be governed by, and need to work within, the framework established by the employer. This should be clearly set down in company policy with the organisation and arrangements for implementation, such as work procedures and safety rules.

The HSE and other bodies, such as the BSI and IEE, have published guidance covering many aspects of EAW which serve as useful information concerning the means for achieving compliance. Many of these publications are referred to in the list of references.

References

1. British Standards Institution, *British Standards Catalogue*, BSI, London (latest edn)
2. Institution of Electrical Engineers, *Requirements for Electrical Installations*, 16th (or latest) edn, IEE, London (BS 7671) (1992)
3. H. M. Government, *The Electricity at Work Regulations 1989*, SI 1989 No. 635, HMSO, London (1989)
4. Health and Safety Executive, *Memorandum of Guidance on the Electricity at Work Regulations 1989*, Health and Safety Series Booklet HS(R)25, HSE Books, Sudbury (1989)
5. Health and Safety Executive, *Approved Code of Practice No. COP 34, The Use of Electricity in Mines*, HSE Books, Sudbury (1989)
6. Health and Safety Executive, *Approved Code of Practice No. COP 35, The Use of Electricity in Quarries*, HSE Books, Sudbury (1989)
7. Hughes, E., *Electrical Technology*, Longmans, London (1978)
8. British Standards Institution, BS 4727, *Glossary of electrotechnical, power, telecommunications, and electronics, BSI, London*
   Part 1 – Terms common to power, telecommunications and electronics
   Part 2 – Terms particular to power engineering
   Part 3 – Terms particular to telecommunications and electronics
9. British Standards Institution, *BS 7430, Code of practice for earthing, BSI, London (1991)*
10. Health and Safety Executive, *Avoidance of Danger from Overhead Electricity Lines*, Guidance Note GS 6, HSE Books, Sudbury (1997)

11. Health and Safety Executive, *Avoiding Danger from Underground Services*, Health and Safety Guidance Booklet No. HS(G)47, HSE Books, Sudbury (1989)
12. Health and Safety Executive, *Electricity at work; safe working practices*, Booklet HS(G) 85, HSE Books, Sudbury (1993)
13. British Standards Institution, *BS EN 60900, Hand tools for live working up to 1000 V ac and 1500 V dc*, BSI, London (1994)
14. British Standards Institution, *BS EN 60903, Gloves and mitts of insulating material for live working*, BSI, London (1993)
15. British Standards Institution, *BS 697, Specification for rubber gloves for electrical purposes*, BSI, London (1986)
16. British Standards Institution, *BS 921, Specification for rubber mats for electrical purposes*, BSI, London (1994)
17. British Standards Institution, *BS 5958, Code of Practice for control of undesirable static electricity*, BSI, London (1991)
18. British Standards Institution, *BS 6742, Electrostatic painting and finishing equipment using flammable materials*, BSI, London (1987/1990)
19. Health and Safety Executive, *Electrical safety on construction sites*, Booklet HS(G) 141, HSE Books, Sudbury (1995)
20. British Standards Institution, *BS EN 60529, Specification for degrees of protection provided by enclosures (IP code)*, BSI, London (1992)
21. British Standards Institution, *BS 7375, Code of practice for distribution of electricity on construction and building sites*, BSI, London (1996)
22. British Standards Institution, *BS 5345, Code of practice for selection, installation and maintenance of electrical apparatus for use in potentially explosive atmospheres (other than mining applications or explosive processing and manufacture)*, BSI, London
23. British Standards Institution, *BS EN 60079–14, Electrical apparatus for explosive gas atmospheres, Part 14, Electrical installations in hazardous areas (other than mines)*, BSI, London (1997)
24. British Standards Institution, *BS 5501, Electrical apparatus for potentially explosive atmospheres*, BSI, London
25. British Standards Institution, *BS 2769, Hand held electric motor-operated tools*, BSI, London
26. British Standards Institution, *BS 4293, Specification for residual current-operated devices*, BSI, London (1993)
27. British Standards Institution, *BS 7071, Specification for portable residual current devices*, BSI, London (1992)
28. Health and Safety Executive, *Maintaining portable and transportable electrical equipment*, Booklet HS(G) 107, HSE Books, Sudbury (1994)
29. Institution of Electrical Engineers, *Code of practice for in-service inspection and testing of electrical equipment*, IEE, London (1994)

Chapter 5

Statutory examination of plant and equipment

J. McMullen

5.1 Introduction

The industrial revolution began in the late 18th century with the mechanisation of the textile industry and subsequent major developments in mining, transport and industrial production, based upon Britain's rich mineral resources such as coal and iron ore, and the use of steam power.

The great industrial towns such as Manchester began to expand with steam power providing the impetus, and this great human exploit was subsequently to make Manchester the world's first industrial city. The industrial conurbation within a 10 mile radius of central Manchester contained in excess of 50 000 boilers – the largest concentration of steam boilers in the world – supplying power to textiles, engineering and to other industries of the period.

The demand by industry for ever-higher boiler operating pressures was outstripping the ability of engineers to meet it safely. As a consequence there was great public concern over boiler explosions which were occurring around Greater Manchester at an alarming rate. The boilers were literally blowing to pieces, causing multiple deaths and injuries.

From 1851 the famous Manchester engineer, Sir William Fairbairn, began arguing for periodic inspection of boiler plant. An advocate of high pressure, in the interests of economy, he decided that the explosions arose from avoidable mechanical causes which could be located in time by carrying out periodic inspections. So in 1854 he enlisted the help of two other eminent Manchester men – Henry Houldsworth, master cotton spinner, and the celebrated engineer, Sir Joseph Whitworth – to form an Association for boiler inspections. This became known as the Manchester Steam Users' Association.

The Boiler Explosions Act 1882 made the reporting of all boiler explosions compulsory and required enquiries to be conducted by Board of Trade surveyors as to the causes and circumstances of the explosions.

At this stage there was still no statutory requirement to have boilers inspected, but, as the 20th century approached, the evidence of over 1000 enquiries held under the Boiler Explosions Act clearly demonstrated the value of regular thorough inspections by competent engineers. The result, included in the Factories and Workshops Act of 1901, was a requirement that steam boilers be subjected to periodic thorough internal examinations by competent persons, and so became the first piece of legislation requiring an item of engineering plant to be inspected, and laid the foundations for the extensive and varied provisions of present day UK law. Electrical equipment inspections and insurance became a prominent part of the insurance portfolio towards the end of the 19th century, with lifts and cranes from early in the 20th century. Dust extraction plant (now referred to as local exhaust ventilation) and power presses followed much later.

5.2 Legislation

The legislation requiring the statutory examination of plant and machinery is undergoing significant changes and is taking a more flexible approach to periodic examinations than earlier prescriptive laws. However, many of the requirements of the latter have been retained in the text in the following sections since they offer guidance on sound and proven inspection procedures. A summary of the principal inspection requirements under existing legislation for pressure systems, lifting and handling equipment, power presses, press brakes, local exhaust ventilation equipment and electrical equipment and installations is contained in *Table 5.1*.

5.3 Pressure systems

Under the Pressure Systems and Transportable Gas Container Regulations 1989 the requirements for examination have become much less prescriptive in that statutory reporting forms no longer need be used and the phasing of the examinations can be related more closely to operating circumstances. Its main requirements are summarised below.

Many of the techniques and practices that developed under the older legislation provide sound safety and engineering guidance and it is sensible to continue using them where they do not clash with the latest requirements. Some of them are described below.

5.3.1 Requirements of the Pressure Systems and Transportable Gas Containers Regulations 1989

These Regulations have been made under HSW and deal with broad objectives to be achieved in the operation of complete pressure systems at all places of work, as opposed to dealing with specific requirements for specific pieces of plant under FA 1961.

Table 5.1 Summary of principal statutory inspection requirements

Statute	*Class of plant*	*Period between examinations – months*
The Factories Act 1961, S. 27	Cranes and lifting machines	14
The Factories Act 1961, S. 22	Hoists and lifts	6
The Factories Act 1961, S. 26	Chains, ropes and tackle	6
The Offices, Shops and Railway Premises (Hoists and Lifts) Regulations 1968		6
Construction (Lifting Operations) Regulations 1961, Reg. 28	Lifting appliances (machines)	14
Construction (Lifting Operations) Regulations 1961, reg. 40	Chains, ropes and tackle	6
Construction (Lifting Operations) Regulations 1961, reg. 46	Hoists	6
Docks Regulations 1988	Lifting machinery	12
Docks Regulations 1988	Lifting tackle	12
Shipbuilding and Ship Repairing Regulations 1960, reg. 34	Lifting appliances (machines)	12
Shipbuilding and Ship Repairing Regulations 1960, reg. 37	Chains, ropes and tackle	6
Quarries (General) Regulations 1956, reg. 13	Cranes and lifting appliances	14
Pressure Systems and Transportable Gas Containers Regulations 1989, reg. 9	All pressure plant	As scheme of examination
Control of Substances Hazardous to Health Regulations 1994, reg. 9	Local exhaust ventilation plant and dust fume extraction plant	1–6
Electricity at Work Regulations 1989 (IEE Regs)	Electrical installations	3–60 depending on application
Lifts Regulations 1997	Applies to the construction of new lifts	
Power Presses Regulations 1965	Power press	6–12 depending on guard type

Notes: LOLER will, when implemented, modify inspection requirements for lifting equipment.

Pressure systems are defined as:

(a) a system comprising one or more pressure vessels of rigid construction, any associated pipework and protective devices;
(b) the pipework with its protective devices to which a transportable gas container is, or is intended to be, connected; or
(c) a pipeline and its protective devices which contain, or are liable to contain, a relevant fluid, but do not include a transportable gas container.

Relevant fluid is defined as:

(a) steam;
(b) any fluid or mixture of fluids which is at a pressure greater than 0.5 bar above atmospheric, and which fluid or mixture of fluids is:
    (i) gas, or
    (ii) a liquid which would have a vapour pressure greater than 0.5 bar above atmospheric pressure when in equilibrium with its vapour at either the actual temperature of the liquid or 17.5 degrees Celsius; or
(c) a gas dissolved under pressure in a solvent contained in a porous substance at ambient temperature and which could be released from the solvent without the application of heat.

The Regulations place obligations on anyone who manufactures or constructs a new pressure system, and anyone who repairs or modifies a new or existing pressure system or part of it, to ensure that no danger will arise when it is operated within the safe operating limits specified for that plant. Regulations 32–36 of FA were revoked when these new Regulations were implemented.

The other main requirements of the Regulations are:

- the user, or owner in the case of a mobile system, must establish the safe operating limits of the system
- the user must have a written scheme of examination for the system
- the user must maintain the system
- the user must have operating instructions and ensure that the system is only operated in accordance with those instructions.

The written scheme of examination is one of the most important features of the new Regulations and is required to be compiled before a pressure system can be operated. Incorporated in the scheme must be details of the pressure vessels, protection devices and pipework where a defect in them could give rise to danger. In addition, it must specify the nature and frequency of examinations and the measures necessary to prepare the system for safe examination over and above the precautions that the user or owner of the system would reasonably be expected to take.

The onus for ensuring that these requirements are met lies with the user of an installed pressure system or the owner or filler of a mobile system – as opposed to the employer or occupier under the FA.

A report of the periodic examination by the competent person must be given to the user or owner of the system within 28 days. However, if there is imminent danger from the continued operation of the system a copy should be provided to the enforcing authority and the report must be provided within 14 days. The Regulations also require that certain records are retained. These include:

- the last report of examination
- any previous reports that required changes to the safe operating limits or repairs
- any documentation provided by the manufacturers.

The requirements of these Regulations regarding a written scheme of examinations and the carrying out of the examinations also apply to existing pressure plant which had previously been examined under the Factories Act 1961.

These Regulations are supported by Approved Codes of Practice[1,2] which are accompanied by Guidance Notes. The Approved Codes of Practice outline, in relative terms, the qualifications to be held by a competent person or authority to enable them to compile a written scheme for a major, intermediate or minor system. Guidance on the frequency of examination of pressure plant can be found in these ACOPs.

5.3.2 Boilers

A definition of a boiler is 'Any closed vessel in which for any purpose steam is generated under pressure greater than atmospheric, and includes any economiser used to heat water being fed to any such vessel, and any superheater used for heating the steam at constant pressure'.

Any reference to a steam boiler also includes all fitments and attachments. These requirements apply to steam raising plant only and not to hot water boilers used in central heating installations.

All existing boilers should have a safe operating limit which for a new boiler is the pressure specified by the manufacturer, but for existing boilers is the pressure specified on the last report of thorough examination. This is the lift-off pressure of the safety valve although it is normal to run the boilers at 90–95% of this to minimise continual leakage from the safety valves.

All boilers are required to have a safety valve so adjusted as to prevent the boiler being worked at a pressure greater than the safe operating limit, a suitable stop valve, steam pressure gauge, at least one water level gauge, means for attaching a test pressure gauge and unless externally fired be provided with a low water alarm device. The nature and frequency of examination of boilers will be specified in the written scheme of examination.

The examination of a steam boiler is normally in two parts:

1 A thorough examination, both internally and externally by a competent person, when it is cold, after the interior and exterior have been

properly prepared in accordance with one or more of the following provisions:

(a) Opening up and cleaning out and scaling of the fire and water sides including the removal of all access doors from manholes, mudholes and handholes.
(b) Opening out for cleaning and inspection of fittings including pressure parts of automatic controls, safety valves and water gauges, and blow down valve.
(c) In the case of water-tube boilers, the removal of drum internal fittings as required by the competent person.
(d) In the case of shell-type boilers, the dismantling of firebridges if made of brick, and furnace protective brickwork at specified periods.
(e) All brickwork, baffles and coverings must be removed for the purpose of the thorough examination, to the extent required by the competent person, but in any case to expose the headers, seams of shells and drums:
    (i) not less frequently than once every six years in the case of steam boilers in the open, or exposed to the weather or damp, and
    (ii) not less frequently than once every ten years in the case of every other steam boiler.

2 The boiler (including economiser and superheater if fitted) must be thoroughly examined by a competent person when it is under normal steam pressure. This examination should be made on the first occasion when the steam is raised after the examination when the boiler is cold, and must consist of ensuring that the safety valve is adjusted so as to prevent the boiler being worked at a pressure greater than the safe operating limit. It should then be locked off to prevent unauthorised tampering with the setting. The operation of the pressure gauges, water gauges, float controls, alarms and cut-outs must also be checked.

Whilst the examination of the boiler, like any other item of statutory plant, is based upon a visual inspection, it may be supplemented by: material thickness measurement, proving a clear waterway through the tubes, withdrawal of sample tubes for evidence of deterioration of wall thickness, non-destructive testing for cracking/flaws by ultrasonic, radiographic, magnetic particle or dye penetrant testing, or by hydraulic testing. Also, as part of the examination the peaking (deviation from circular shape) at the longitudinal seam should be measured. The periods between ultrasonic examination of the seam can be determined from information in the SAFed guidance on the examination of longitudinal seams[3]. For certain boilers with flat end plates it is normal to undertake ultrasonic examination of the end plate welds to detect any cracking from the water side of the weld. Guidance on the nature and frequency of these inspections can be found in SAFed publications[4,5].

Entry into any boiler, which is one of a range of two or more, is forbidden unless all inlet pipes through which steam or hot water could enter the boiler have been disconnected, or the valves controlling entry of steam or hot water have been closed and locked. Particular attention should be paid to blow-down valves which discharge into a common line or sump, where special safety procedures must be followed[6]. Suitable precautions should also be taken to ensure that the boiler is free from dangerous fumes and has adequate ventilation.

The purpose of the examination is to ascertain the material condition of the boiler with particular reference to any defects which could affect the continued safe working of the boiler at its current safe operating limits. During the examination when cold the competent person will be checking for the following types of typical defects:

1 *External* – Wastage from corrosion due to leakage at joints between fittings and shell, tubes and tubeplates, or at seams in riveted boilers; wastage of manhole, mudhole and handhole joint seatings; general physical damage.
2 *Fire side* – Wastage in combustion chamber plates due to leaking tubes, stay tubes or stays; furnace flame impingement damage; erosion of material by fast moving gases and/or entrained particles – particularly where gases are hottest and changing direction (i.e. combustion chamber ends of tubes in horizontal shell-type boilers); overheating damage due to scale or sludge build-up, or water shortage; furnace bulges and combustion chamber crown bulges.
3 *Water side* – Chemical damage causing wastage, pitting, thinning and necking of material due to oxygen or other chemical corrosion; mechanical damage caused by expansion and contraction, and grooving at plate flange bends, evidence of 'peaking' at the longitudinal weld caused during the manufacturing process, cracking at the tube/tubeplate joints, and other slowly developing fatigue cracks at a variety of locations.

Examples of typical defects are shown in *Figure 5.1(a)* and *(b)*.

Reports which should cover both parts of the examination should be issued by the competent person. Electronic storage of data is acceptable subject to suitable security arrangements for access to it and the facility to produce a hard copy if required. A list of the items that should be included in the report is given in Approved Codes of Practice[1,2]. For smaller users who wish to use a standard form of report, a suitable form is available from HSE Books.

When the examination reveals defects which affect the safe working of the boiler at the current safe working limit, the report must make recommendations on conditions to be met before the boiler can be returned to operation. The conditions may include certain defined repairs, or the reduction of the safe working limit. In both cases a copy of the report of examination must be forwarded to the enforcing authority within 14 days.

Where a defect has been found in the boiler and subsequent repairs carried out, the competent person should satisfy himself that the repair work has in fact been carried out satisfactorily before the boiler is released for service. This will normally include review of material certificates for new components, review of weld procedures, welders qualification and NDT reports. Additionally the competent person will examine the boiler

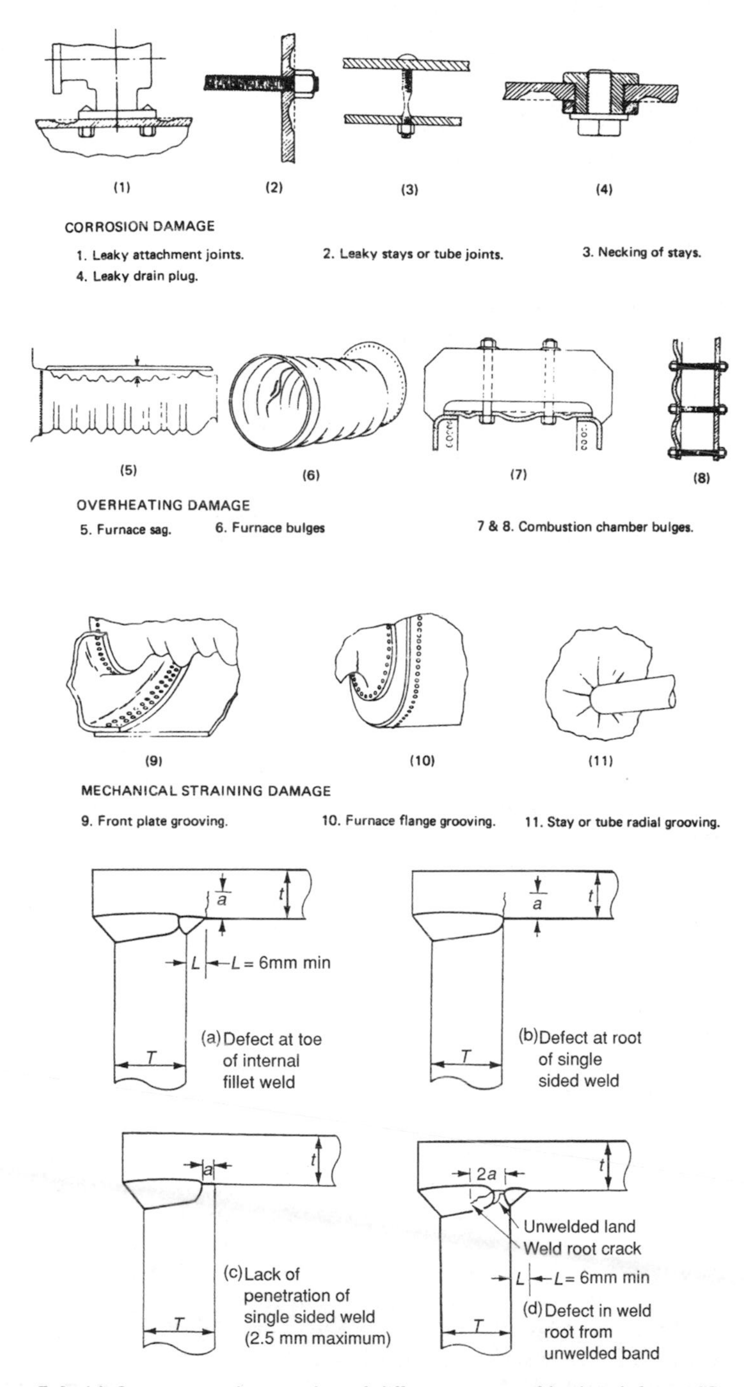

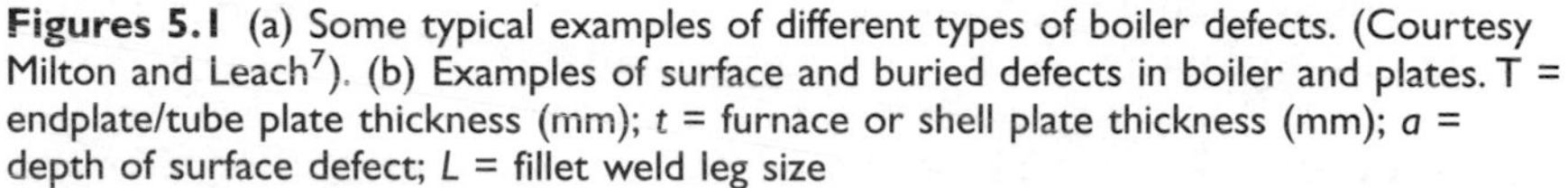

Figures 5.1 (a) Some typical examples of different types of boiler defects. (Courtesy Milton and Leach[7]). (b) Examples of surface and buried defects in boiler and plates. T = endplate/tube plate thickness (mm); t = furnace or shell plate thickness (mm); a = depth of surface defect; L = fillet weld leg size

during repair and witness a hydraulic test before it is returned to service.

5.3.3 Steam and air receivers

These may be defined as:

1 *Steam receiver* – Any vessel or apparatus (other than a steam boiler, steam container, steam pipe or coil, or part of a prime mover) used for containing steam under pressure greater than atmospheric pressure.
2 *Steam container* – Any vessel (other than a steam pipe or coil) constructed with a permanent outlet into the atmosphere, and through which steam can be passed at approximately atmospheric pressure, for the purpose of heating, boiling, drying, evaporating or other similar purpose.
3 *Air receiver* – Any vessel for containing compressed air and connected to an air compressing plant, a fixed vessel for containing compressed air or compressed exhaust gases used for the purpose of starting an internal combustion engine, a fixed or portable vessel used for the purpose of spraying by means of compressed air, any paint, varnish, lacquer or similar substance, or any vessel in which oil is stored and from which is forced by compressed air.

There are other types of vessels and tanks, not covered by the definition of air receivers, from which fluids are blown by compressed air, and whilst non-statutory at present, it is advisable to subject such vessels to similar inspection procedures.

The safe operating limit is a legal term which applies to pressure vessels including steam receivers and air receivers. In the case of a new receiver, it means the pressure specified by the manufacturer, and in the case of a receiver that has been examined in accordance with the written scheme of examination, that specified by the competent person.

Both the steam receiver and the air receiver should be provided with a suitable reducing valve or other automatic appliance to prevent the safe operating limit being exceeded. If the vessel is not suitable for the pressure of the source of supply, a pressure gauge, a safety valve with a vent to atmosphere and, where more than one vessel is in use in the factory, each must bear a separate distinguishing mark. In the case of the steam receiver only, it must have in addition a suitable stop valve. Air receivers must be fitted with a drain plug, provided with a suitable manhole, handhole or other means which will allow the interior to be thoroughly cleaned and have the safe working pressure marked on it.

The frequency of examination of this equipment will be specified in the written scheme but most steam and air receivers will have a frequency of examination of 26 months. If a vessel is so constructed that the internal surface cannot be thoroughly examined, then either non-destructive testing or a hydraulic test may be carried out.

As with boilers the scheme of examination for these vessels should be drawn up by a competent person before they are put into service in a system. The manner of the examination is left to the competent person carrying it out.

When making his report, the competent person will need to have examined the vessels internally and externally, checked that the required fittings are provided, properly adjusted and in good working order and that the vessels are clearly marked with an identification number and the safe working pressure where necessary. A report of the examination should be sent to the user or owner within 28 days.

Where a defect is found, a copy of the report should be sent to the enforcing authority within 14 days.

The most common defect found in these types of vessels is thinning or pitting of the shell plates due to corrosion on the inner surface and this is normally countered by reducing the safe operating limit, which is calculated from plate thicknesses and other data ascertained at the examination.

5.4 Lifting and handling plant

The range of what has generically become known as lifting and handling plant has expanded over the years in both scope and sophistication, with many innovative designs of plant now being used in industry and for various handling processes. Because of the very nature of this type of plant, together with its versatility, it is used in many industrial locations, and as such a number of Acts and Regulations make provision for it. These include: the Factories Act 1961; the Offices, Shops and Railway Premises (Hoists and Lifts) Regulations 1968; the Miscellaneous Mines Order 1956; the Quarries Order 1956; the Shipbuilding and Ship Repairing Regulations 1960; the Docks Regulations 1988; the Loading and Unloading of Fishing Vessels Regulations 1988, and the Construction (Lifting Operations) Regulations 1961.

Under these provisions a report has to be made following each examination and the possibility of rationalising the many reporting forms used for lifting plant has been reviewed with the object of reducing the number of report forms and test certificates, and so simplifying the administrative procedures. Reporting requirements are contained in the Lifting Plant and Equipment (Record of Test and Examination etc.) Regulations 1992.

The above legislation contains differing requirements for similar machines and the definition of lifting equipment may vary between them. For this reason, and to implement certain requirements of the amended Work Equipment Directive[8], the HSE have formulated generic regulations for lifting equipment which are known as the Lifting Operations and Lifting Equipment Regulations 1998 (LOLER). These are intended to be implemented late in 1998. New lifting equipment supplied after 1 January 1993 is subject to SMSR.

5.4.1 Lifts and hoists

There are many variations on the fundamental design concepts for hoists and lifts, but basically they are of four main classes: (1) electric, incorporating winding gear and a traction sheave drive or winding drum, (2) hydraulic, either direct or indirect acting, (3) manual service type, and (4) continuous paternoster type lifts. Each may be classified as passenger or goods lifts only, or a combination of both. Continuous lifts (known as paternosters)[9] have no doors and the passengers step into and out of the car as it passes the landing. These are normally to be found in hospitals and colleges where the rapid mobility of staff and students is required. Due to a number of serious incidents the HSE have prohibited installation of paternosters within premises subject to HSW.

The earlier legislation covering the construction, use and inspection of hoists and lifts is ss. 22–25 of FA 1961 and the Offices, Shops and Railway Premises (Hoists and Lifts) Regulations 1968 (OSRP Regs). However, these are being overtaken by the Lifts Regulations 1997 for the construction of new lifts and hoists, and by LOLER concerning their use and examination.

A hoist or lift is defined as a machine that incorporates a platform or cage where the direction of movement is restricted by a guide or guides. In the case of existing service lifts, in addition to the above, the floor area of the cage should be no more than 1.25 m^2 and its maximum height 1.2 m. A greater height is allowed if the floor area is subdivided into compartments.

In-service requirements for existing lifts are that every hoist or lift should be of good mechanical construction, sound material and adequate strength, and should be properly maintained. In the case of lifts constructed after 1937 in factory premises, and after 1968 in offices, the gates of the lift must be fitted with interlocks, usually of the electro-mechanical pre-locking type and of fail safe design, so that the landing gate cannot be opened unless the cage or platform is at the landing, and that the cage or platform cannot be moved unless the gates are closed and locked. An electrical interlock is required on the cage gate such that the lift cannot be moved when the cage gate is open.

This provision does not apply to manual service lifts or continuous lifts. Where the hoist or lift is used for carrying persons, as opposed to 'goods only' lifts, it must be fitted with: (a) an automatic limiting device to prevent the cage or platform over-running, (b) at least two suspension ropes or chains separately anchored, each capable of taking the maximum working load and weight of the cage etc., and (c) safety gear that will support the cage with its maximum working load in the event of the suspension ropes failing. Under the Building Regulations 1985, the enclosing shaft of all lifts must be of a fire-proof construction.

The frequency of thorough examinations for a hoist or lift by a competent person under the FA and OSRP Regulations is six months. In the case of continuous lifts and manual service lifts the frequency is every 12 months. The results of examinations must be recorded and given to the employer or owner of the equipment, but there is no specific form for reporting. However, the data to be recorded are listed in schedules to Regulations[10] and an HSE guidance document[11]. Where defects are found necessitating repairs to ensure continuing safe operation, a copy of the examination report must be sent to the enforcing authority within 28 days.

Under the Hoists Exemption Order 1962, a number of types of hoist or lift are exempted from certain of the Regulations, depending upon the design. The main examples are: lifts descending from pavement level to the basement of a building, mobile and fixed lifts used for loading and unloading goods directly from vehicles where the maximum height of travel is restricted, electrical and mechanical service lifts where either the landing and cage entrances are fitted with lattice type gates or the lift cage has no door and moves in a flush lined shaft, and hoists used solely for lifting materials directly onto a machine.

The purpose of the examination is to ascertain whether the lift can continue to be used with safety up to its designated operating limit for the period of time until the next inspection and, if not, to specify any repairs, renewals or adjustments necessary to ensure this. The person making the examination must consider all factors affecting safety including the mechanical and electrical components of the installation.

The safety of passenger lifts relies on the proper operation of many components in the installation but in the case of an electrical lift a Guidance Note[12] contains advice for examination and testing of specific components. This guidance is undergoing revision and a new version is expected to be published in 1999.

There is no legal requirement to test a lift prior to commissioning, although clients may require the competent person to witness the pre-commissioning tests carried out by the manufacturers or contractors. The Lifts Regulations 1997 will require every passenger lift to be tested following installation after 1 July 1999. The types of tests that would normally be carried out will be included in an EN standard[13].

5.4.2 Cranes and lifting machines

A vast range of cranes and lifting machines are in current use, with the following categories being the most commonly encountered: tower cranes, overhead travelling cranes, mobile cranes, portal cranes, derrick cranes and goliath cranes; and in the case of lifting machines: electric hoist-blocks, manual pulley or chain blocks, fork trucks, swinging jibs and electric teagle hoists.

Lifting machines are defined differently, depending upon the industry and which particular piece of legislation applies. For example, s. 27 of the FA defines a crane as a 'lifting machine', and includes in the definition a crab, winch, teagle, lifting block (manual or electric), gin wheel, transporter or runway (sometimes referred to as a trolley). The Construction (Lifting Operations) Regulations 1961 define a 'lifting appliance' as including shear legs, excavator, dragline, piling frame, aerial ropeway or overhead runway in addition to the above, and under the Shipbuilding and Ship Repairing Regulations 1960 a 'lifting appliance' also includes a derrick. The Miscellaneous Mines Order and the Quarries Order define a 'lifting machine' as simply a crane, crab or winch. Finally, the Docks Regulations 1988 define the 'lifting appliance' more generally as a stationary or mobile appliance, including anchorages, fixings and supports when used on dock premises for suspending, raising or lowering or moving loads, and so includes fork trucks, chain blocks, lifting cradles, cranes, straddle carriers and access lifting platforms. New lifting equipment for goods is required to comply with the essential safety requirements of SMSR.

It should be noted that the scope of the new Docks Regulations is wider than the revoked Docks Regulations 1934, and now refers to all places where 'dock operations' are carried out including neighbouring land and buildings upon the land, used for landing, storage, sorting, inspection, checking, weighing or handling of the goods.

Whatever the strict legal definition of the lifting machine, all parts of the fixed or mobile working gears, including anchorages, must be of good construction, sound material, adequate strength, free from patent defect and properly maintained. The Loading and Unloading of Fishing Vessels Regulations 1988 and the Docks Regulations 1988 also imply that it must be of good design and fit for the purpose for which it is to be used.

A new lifting machine must have a Declaration of Conformity, be CE marked and be tested and thoroughly examined by a competent person and have its safe working load (above 1 tonne in the case of the Construction Regulations) marked on it before being taken into use for the first time. In the case of the Docks Regulations 1988 and the Loading and Unloading of Fishing Vessels Regulations 1988, an inspection and test is required for plant that has been modified, altered and also following a major repair that is likely to affect its strength or stability. Under the Construction Regulations, this testing requirement is rather more complex. Generally, any crane, crab or winch must be tested and thoroughly examined every four years, and all pulley blocks, gin wheels and shear legs should be tested before use. In addition to this, every time a crane is erected on site, or after partial dismantling and re-erection on another part of the site, removal or adjustment involving the anchorage and/or ballast, also any major repairs, or modifications to members in the direct line of stress of a crane, crab or winch and affecting its strength or stability, or any substantial alteration to a pulley block, gin wheel, shear legs greater than 1 tonne, it must be tested and thoroughly examined. This provision to test a crane after erection on site would apply to a tower crane or very large mobile crane of the maxi-lift type but not to the normal size of mobile cranes. As it is unlikely that such cranes will be on site for four years, these cranes will be tested at a greater frequency. In the case of a lifting appliance on board ship, under the Docks Regulations, this must have been tested within the last five years.

All lifting appliances must be thoroughly examined by a competent person every 14 months under the Factories Act, the Construction Regulations, the Miscellaneous Mines and Quarries Orders, and every 12 months under the Shipbuilding and Ship Repairing Regulations, the Loading and Unloading of Fishing Vessels Regulations and the Docks Regulations. In addition, the Docks Regulations stipulate that more frequent inspections can be specified by the competent person when deemed necessary. Also, under the Miscellaneous Mines and Quarries Orders a crane, grab or winch, if once dismantled or out of regular use for more than two months, should be inspected by a competent person before going back into use. Moreover, the Loading and Unloading of

Fishing Vessels Regulations require the lifting appliance to be visually inspected by a responsible person before loading/unloading commences, and the Construction Regulations require all lifting appliances to be inspected by the operator at least once every week and the results reported on a prescribed form.

The Report of thorough examination must be on a form containing prescribed particulars. Where the machine has a defect and cannot continue to be used with safety, then a copy of the report must be sent to the Health and Safety Executive within 28 days. The Docks Regulations also stipulate that the reports are to be retained for at least two years; this practice is, however, recommended in all cases, regardless of which legislation is applicable.

The provision for establishing the maximum safe working load that can be lifted at a particular jib radius of the crane has become a complex issue under various pieces of legislation. Under the Factories Act, a mobile crane with a varying jib radius must be fitted with a safe load indicator, or a table of safe working loads used in conjunction with a pendulum type indicator mounted on the jib such that it is clearly visible to the operator and shows the radius of the crane hook. Again the Shipbuilding and Ship Repairing Regulations simply require a radius indicator for cranes with derricking jibs with corresponding indication of safe working loads. The Construction Regulations are slightly more demanding in the sense that all mobile, tower and derricking cranes above 1 tonne must be fitted with an approved type of automatic safe load indicator. Where this safe load indicator has been wholly or partially dismantled it must be tested by a competent person before the crane is taken back into service and the results reported on a prescribed form. The requirements under the Docks Regulations are somewhat more onerous in the sense that they specify three separate situations for various types of plant. Firstly, a load/radius indicator is required for such as tower/cantilever cranes of the hammer-head type; secondly, a load/angle indicator for mobile cranes operating with fly jib or locked manual extension; and thirdly, an approved automatic safe load indicator for mobile cranes where the safe working load is greater than 1 tonne. SMSR requires all lifting machinery with a maximum working load of 1000 kg or more to be fitted with devices which will warn the driver and prevent dangerous movements of the machinery in the event of overload or movements due to the safe working load being exceeded or if movements conducive to overturning are initiated.

The Construction Regulations and the Docks Regulations state that the safe working load is that load specified in the latest certificate of test obtained from the manufacturers or suppliers, or the latest report of thorough examination. This latter point is relevant when the competent person has derated the lifting machine owing to certain mitigating circumstances. It should be noted at this stage that, whilst not quoted in the legislation, the safe working load (SWL) will in most cases be the same as the working load limit (WLL). However, if a lifting machine has been derated because of the heavily corrosive environment in which the equipment is being used, then the SWL stated on the equipment will be less than the WLL depending upon the amount of derating, thus building into the equipment a further factor of safety.

The philosophy underlying any inspection of lifting machines should entail a check on all items in the direct line of stress. Commencing at the hook this would comprise the following major components, where appropriate, but is by no means exhaustive:

1 Hook return block, jib head pulleys, lifting rope and rope anchorage points and fixings. Condition of lifting ropes, bridle ropes, compensating pulleys, wedge type socket fixings and bulldog grip fixings.
2 Jib and mast structures. General condition of main structural members (chords), bracings, struts, ties, jib foot pin, in terms of deformation and impact damage, and a visual inspection of the welds.
3 Crane chassis structure, including the 'A' frame in the case of a mobile crane and loose or failed bolts in the crane slewing ring.
4 Machine winding gear.
5 Control, including any derricking limits etc.; particular attention being paid to any interlocking arrangements between separate control functions.
6 Safe working load indicators. This can vary from a simple notice of SWL's and pendulum-type load/radius indicators on the jib, through to a dynamometer incorporating a load cell on the jib with a display readout in the operator's cab, to sophisticated microprocessors incorporating automatic safe load indication and overload alarms. The competent person must satisfy himself that the type of safe load indicator fitted conforms to the relevant statutory requirements, and that the indicator is accurate for the configuration of the crane rigging.

Generally, the testing procedure would comprise an initial inspection of the lifting machine, a functional test during full operation of the crane, application of the safe working load and the proof load, a measurement of deflections of the structure (where appropriate), and a re-inspection to ensure permanent deformation has not taken place.

In the case of mobile cranes, the test load depends upon the date of manufacture of the crane. For those manufactured before March 1981, the amount the test load exceeds the SWL depends upon the size of crane:

1 Up to 20 tonnes, the test load is the SWL plus 25%.
2 From 20 up to 50 tonnes, the test load is the SWL plus 5 tonnes.
3 Over 50 tonnes, the test load is the SWL plus 10%.

For mobile cranes manufactured after March 1981, regardless of size, the crane must be subjected to a dynamic test with a load of the SWL × 1.1, where the crane is moved through all its motions, plus a static test of the SWL × 1.25 where the load is simply raised slightly off the ground.

Additional special tests are required in the case of cranes with manual extensions and for fly jib extensions.

Under the Construction Regulations the safe load indicator also requires to be tested and reported upon and, in addition, in the case of a tower crane an anchorage test needs to be carried to and reported upon.

Higher proof loads should be applied for stability tests on cranes, but in practice would very rarely be carried out.

For new machinery, SMSR requires a dynamic test of 1.1 × SWL and a static test of 1.25 × SWL unless a relevant standard requires a different test regime.

5.4.3 Miscellaneous lifting equipment

Lifting tackle is variously described as chains, ropes, rings, hooks, shackles and swivels (FA s. 26) plus links, plate clamps and eyebolts (Construction Regulations and Shipbuilding and Ship Repairing Regulations). The Docks regulations refer to *lifting gear* as any gear by which a load can be attached to a lifting appliance but which is not part of either the appliance or the load. SMSR defines *lifting accessories* as components or equipment not attached to the lifting machine but placed between it and the load in order to attach it.

Again the equipment must be of good construction, sound material, adequate strength and free from patent defect. The Loading and Unloading of Fishing Vessels Regulations and the Docks Regulations extend this to include good design, properly installed and assembled, and properly maintained. All items must be separately identified with the safe working load clearly marked, and in the case of multi-legged slings under the Factories Act a table of safe working loads displayed in the distribution store.

The safe working load means that specified on the last test certificate, or report of examination, or as specified by the manufacturer/supplier. The equipment must be supplied with a test certificate or, for lifting equipment subject to SMSR, a Declaration of Conformity, before being used under any of the relevant Statutory Instruments, and re-tested after any substantial repair or modification.

In the case of the Factories Act and Construction Regulations, lifting tackle must be thoroughly examined every six months by a competent person. Under the Shipbuilding and Ship Repairing Regulations, lifting gear and chains must also be examined every six months, but in the case of wire ropes the frequency is every three months; where the examination of the rope reveals that a wire strand of the rope or rope sling has broken, then the examination frequency must be monthly until such time as the rope is discarded. The Loading and Unloading of Fishing Vessels Regulations and the Docks Regulations state 12 monthly examination frequency, or a shorter period as specified by the competent person; the more frequent examination could be carried out by a responsible person on site.

Chains, links and shackles of wrought iron construction (where still in use) must be annealed every 14 months to comply with the Factories Act, Construction Regulations and the Shipbuilding and Ship Repairing Regulations; where they are manufactured from 0.5 inch bar or smaller the frequency is increased to six monthly.

Under the Construction Regulations, hooks used on cranes must be fitted with a safety catch, or the hook must be of such a shape as to prevent the possibility of the sling becoming displaced.

Some of the Statutory Instruments such as the Construction Regulations and the Shipbuilding and Ship Repairing Regulations outline certain discard-criteria for wire ropes, such that if in any length of 10 diameters the total number of visible broken wires exceeds 5% of the total number of wires in the rope, then the rope should be rejected.

The proof load parameters for testing lifting tackle are complex, particularly with regard to the Shipbuilding and Ship Repairing Regulations, and the Docks Regulations, and the reader is referred to the Regulations for further details.

Regulations[10] no longer demand the use of a pro-forma for recording the results of examinations but give details of what is required regarding the content of the reports while allowing flexibility in the form used. Reports should be kept for at least two years or such longer period that particular legislation may require.

In addition, under these Regulations where a defect is found in the equipment, a copy of the report does not need to be sent to the Health and Safety Executive.

Lifting tackle is not only used as ancillary equipment to a lifting operation, but will also be an integral part of the lifting machine, such as lifting ropes, hooks etc. on a statutory item such as a crane, or a non-statutory item such as a fork-lift truck or a motor vehicle lifting table. In the latter cases, even though the items are non-statutory, the ropes/chains are statutory items and require to be inspected every six months.

The examination is to assess the suitability of the equipment for continued use, and the following are some typical defects of lifting tackle: stretching, distortion, corrosion, fatigue cracking, wear between adjacent chain links, broken wires in ropes and degradation of the internal fibre core of the rope etc. The degree of 'acceptable damage' is a matter of judgement on the part of the competent person in many respects, as the dividing line between a defect and an observation can be extremely arbitrary. Also, lifting tackle may appear to be relatively simple items of engineering equipment and, as such, abused in many instances. Notwithstanding this, the stresses imposed on an item of lifting equipment during normal use are highly complex, so it is constructed with substantial factors of safety to withstand shock loads and maltreatment. In addition, wire ropes are a subject in themselves and generally have a factor of safety of about 5:1. However, they can deteriorate rapidly if abused, or if left outside in sub-zero temperatures, or subjected to elevated temperatures from a manufacturing process, or used over salt baths etc.

5.4.4 The Lifting Plant and Equipment (Records of Test and Examination) Regulations 1992[10]

These regulations are not concerned with the actual tests and examinations of lifting items but with revoking all references to 'register', 'certificate' and 'report' in earlier legislation. In addition, they revoke the requirement to report on a 'prescribed form' and also remove any requirement to provide a written report except where required to do so by an HSE inspector. Instead, they provide for the keeping of a 'record' in a form agreed by both the competent person carrying out the test/examination and the plant owner. This enables future records to be stored, transmitted and retrieved by electronic means.

Under these Regulations, reports of examinations must contain the following prescribed particulars:

- Desciption of item, identification mark and location.
- Date and number of the record of the last thorough examination.
- Safe working load or loads and corresponding radii.
- Date of most recent test and thorough examination.
- Details of any defects found and, where appropriate, a statement of the time by which each defect should be rectified.
- Date of completion of thorough examination.
- A declaration that the information is correct and that the equipment has been thoroughly examined and is free from any defects, other than those recorded, likely to affect safety.
- Name and address of the owner of the equipment.
- Name and address of the person responsible for the thorough examination.
- Date the record of the thorough examination is made.
- Name and address of the person who authenticates the record.
- A number or other means of identifying the record.

The detailed information demanded by the Regulations depends upon the circumstances of the examination and reflects the data which should be collected if that examination is to be effective in ensuring the future safe operation of the plant. The four circumstances are:

- Where an examination and a test have been carried out.
- After a period of examination.
- The testing and examination of chains, ropes and lifting tackle.
- Where a copy of the record has to be sent to another person, i.e. if the equipment has to be repaired before being returned to service.

5.4.5 Lifting Operations and Lifting Equipment Regulations 1998

These new Regulations, due to be implemented in December 1998, will revoke or repeal large sections of existing legislation that apply to lifting equipment. LOLER will provide a single set of modern goal setting regulations for lifting equipment which will remove the anomalies of the existing sector specific legislation. It will apply across all industries including those where there is at present no specific lifting law. It will replace the existing pre-HSW lifting legislation and create a simple regime for lifting operations and lifting equipment. It will be risk based.

The proposals for LOLER include an option for routine periodic inspections to be based on either a scheme drawn up by a competent person to match the operating conditions met or examination at prescribed maximum intervals of:

- 6 months for lifting equipment and accessories involved in lifting people
- 12 months for other lifting equipment.

Under these proposals, where an inspection reveals a defect which is, or could become, a danger to persons, the competent person must send a copy of his report to the relevant enforcing authority. This requirement will apply to all types of lifting equipment. An Approved Code of Practice is to be issued to support these new Regulations which will contain guidance and interpretation to enable those responsible for the provision and use of lifting equipment to achieve the goals set by the Regulations.

5.5 Power presses and press brakes

Power presses and press brakes are complex pieces of equipment that have the potential to cause serious injury. For this reason it is important to ensure safe systems of work are in place and that power presses and their associated guarding arrangements are subjected to regular examinations and testing to ensure that all systems are operating properly and have not suffered damage likely to reduce the required level of safety. The Power Press Regulations 1965 provide for such examinations to be carried out at prescribed intervals by competent persons.

The Regulations define a power press as being a press or press brake which in either case is used wholly or partly for the working of cold metal by means of tools or for the purpose of die proving is power driven and embodies a flywheel and clutch. Hydraulic and pneumatic presses are beyond the scope of these Regulations.

Following the introduction of the Regulations in 1965 a number of certificates were issued by HM Chief Inspector of Factories which exempted from the requirements certain classes of power presses including those used for compacting metal powders, presses with a stroke not exceeding 6 mm, turret punch presses and other machinery such as combination machines used solely for punching, shearing and cropping. Other machines beyond the scope of these Regulations include guillotines, riveting machines and special purpose machines such as those used for the manufacture of zip fasteners, eyelets, and similar components.

The main thrust of the regulations is towards the thorough examination and testing of machines and guarding systems before the machine is first taken into use and thereafter at periodic intervals depending on the nature of the tool guarding arrangements. Where the dangerous tool trapping area is protected exclusively by fixed guards the interval between thorough examinations is 12 months, but in all other cases including where interlock guards, automatic guards or photo-electric safety systems are used, the interval is 6 months. The examination requires close scrutiny of the safety critical parts such as the clutch and brake mechanisms and will, at some stage, require dismantling of the flywheel assembly in order to expose the condition of the inner components. The need for dismantling is confirmed in HSE guidance[14].

The results of the thorough examination and test must be entered on an approved form F2197 which must be kept for at least 2 years. Any defects found that may affect safety at the tools must be notified immediately in writing to the user and a copy of F2197 sent to the HSE. Otherwise the user must be provided with a copy of the report within 14 days of the examination.

The setting, resetting, adjusting or trying out of tools on a power press and the proving of the guarding arrangements may only be carried out by someone who:

- is 18 years old or more
- has been suitably trained
- is competent
- has been appointed by the employer in writing.

He would normally be the tool setter.

In addition, the tool guards and safety devices must be inspected and tested during the first 4 hours of a shift by the setter who must sign a register or certificate, which should be kept on or near the press, to verify that the guards are in position and in good working order. Such registers must be kept until 6 months after the date of the last entry.

Although hydraulic press brakes are not included in the Regulations, in their publication on press brakes[15] the HSE recommend that the presses and their associated guarding systems should be treated the same as power presses and subjected to a thorough examination every 6 months in addition to the checks carried out on each shift.

Although the Power Press Regulations are likely to fall victim to the revision of PUWER, it is probable that the core requirements will be retained in the new proposals.

5.6 Local exhaust ventilation

Local exhaust ventilation (LEV) equipment is intended to control mechanically the emission of contaminants such as dust and fumes that are given off during a manufacturing process or in a chemical laboratory. Normally this is done as close to the point of emission as possible using a stream of air to remove the airborne particulate matter and transport it to where it can be safely collected for ultimate disposal. The physical layout and setting of LEV equipment is critical for it to work effectively, and comparatively minor alterations can affect its performance. It is, therefore, important that LEV equipment[16] should be properly designed, manufactured, installed, operated and maintained.

The main elements of the equipment would normally comprise:

1 captor hood,
2 exhaust ducting,
3 extraction fan, and
4 filter/collecting bags.

In addition, the following items would be regarded as LEV equipment:

5 parts of machinery such as integral machine casing or guarding which has a dual purpose of controlling and venting emissions from the process to atmosphere;
6 vacuum cleaners when permanently connected to an exhaust system and fitted to portable tools;
7 flues from a furnace/oven when the plant is producing hazardous or toxic fumes – but not when the flue only serves to create a draught for a combustion process;
8 fume cupboards in chemical laboratories; and
9 low volume/high velocity (lv/hv) extraction for cutting processes.

The major legislative requirements concerning the examination and testing of LEV plant are contained in the Control of Substances Hazardous to Health Regulations 1994 (COSHH) and its supporting Approved Codes of Practice[17,18].

Where an assessment has identified an exposure level above the occupational exposure standard (OES) or maximum exposure limit (MEL) for the substance, then under a hierarchy of preferred controls, LEV equipment must be installed wherever practicable, as opposed to providing personal protective equipment, as the means for reducing the employee's exposure. Such equipment must be properly used by the operator and visually inspected for obvious defects. Further, the employer must ensure that the equipment is maintained and the statutory inspections are carried out. As a part of this inspection it may be necessary to monitor the working environment by air sampling to ensure that the plant is continuing to operate effectively.

COSHH lists, in schedule 3, the frequency of examination of LEV as:

1 All LEVs other than those specific processes mentioned below – every 14 months.
2 Processes in which blasting is carried out in, or incidental to, the cleaning of metal castings, in connection with their manufacture – monthly.
   (It must be noted that these monthly inspections refer to castings only. In the case of blasting other metallic items the inspection frequency is 6 months.)
3 Processes, other than wet processes, in which metal articles (other than gold, platinum, or iridium) are ground, abraded or polished using mechanical power, in any room for more than 12 hours in any week – every 6 months.
4 Processes giving off dust or fumes in which non-ferrous metal castings are produced – every 6 months.
5 Jute cloth manufacture – every month.

Where plant is old, the frequency of inspections may need to be increased.

COSHH does not extend to include asbestos or lead which are covered by extant regulations.

The Control of Asbestos at Work Regulations 1987 requires, *inter alia*, that every employer shall take control measures necessary to protect the health of employees from inhaling asbestos. Details of the pressure and air velocity tests and inspection procedures to be used are given in an Approved Code of Practice[19] which provides for three levels of inspection, i.e. weekly inspections by a responsible person, an examination of new and substantially modified plant (referred to as Part 1 examination), and six monthly inspections by a competent person (referred to as Part II examination).

The competent person may enlist the assistance of specialists to carry out certain tests, but he must make a report within 14 days of the examination. The report of Part I tests should include technical details of the plant.

The Control of Lead at Work Regulations 1980 requires employers to provide such control measures, other than by the use of respiratory equipment or protective clothing, as will prevent the exposure of his employees to lead. An Approved Code of Practice[20] outlines the weekly inspections to be carried out by a responsible person and requires that the annual examination and test covers the condition of the LEV plant, static pressures and air velocities at various points and a check that the lead dust or fume is being effectively controlled. Although the Code does not require it, it would be prudent for the annual examination to be carried out by a competent person.

Apart from the visual inspection of the LEV equipment, a statutory thorough examination under COSHH would include some of the following tests:

1 Tyndall (dust) lamp (*Figure 5.2*) and/or fuming sulphuric acid test;
2 static pressure behind the captor hood. This is perhaps the most important test as this will determine whether the performance of the LEV equipment has altered;

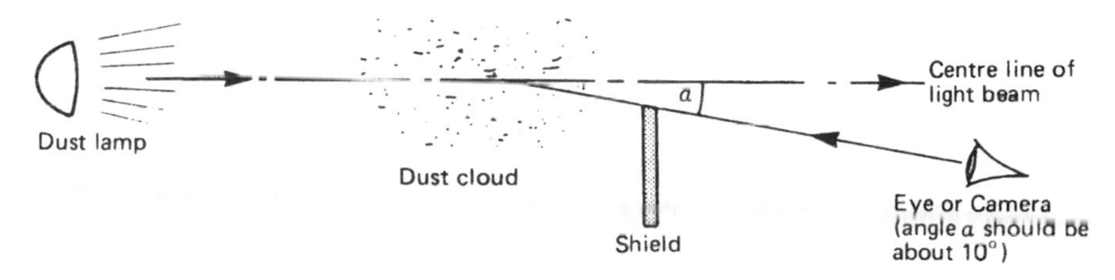

Figure 5.2 Use of dust lamp to see or photograph dust

3 air velocities across the face of the captor hood;
4 centreline velocity of the small duct between the captor hood and the main exhaust ducting;
5 velocity and static pressure at the main exhaust ducting;
6 static and total pressures at the inlet and outlet of the exhaust fan; and

7 static pressure at the inlet and outlet of the dust collector and filters to obtain the differential pressure – particularly required for systems that recirculate the filtered air back into the workplace.

5.7 Electrical equipment and installations

The periodic inspection and testing by a competent person of electrical equipment and installations as a statutory requirement arises only under the Cinematograph (Safety) Regulations 1955 which require inspection and testing by a competent person every 12 months.

However, enforcing authorities may impose licensing conditions requiring petrol dispensing and storage (under the Petroleum (Consolidation) Act 1928), launderettes, sports grounds and buildings used for public entertainment, to be inspected, tested and certified periodically (see also below). Enforcing authority licensing conditions specify annual inspections, testing and certification that the installations comply with the current (16th) edition of the Institution of Electrical Engineers (IEE) Regulations[21], now incorporated into BS 7671[22].

In Scotland compliance with the IEE Regulations is a requirement of the Building (Scotland) Act 1959.

The Electricity at Work Regulations 1989 apply to electrical systems in all work situations. They refer to design parameters such as the circuit strength and capability, insulation, earthing, protective devices and the precautions to be taken when working on electrical equipment. All systems must be maintained, as far as is reasonably practicable, so as to prevent danger. A technique to ensure proper maintenance is through programmed preventive maintenance of which inspection and testing are essential parts. The frequency of maintenance is not specified, but practical experience should be a useful indicator. Factors to consider when determining the frequency of maintenance include mechanical wear and tear, impact damage, corrosion, excessive electrical loading, ageing and environmental conditions.

The Inspection and Testing Guidance Notes to the IEE Regulations suggest the following periods between inspections:

domestic premises	10 years
commercial premises	5 years
educational premises	5 years
hospitals	5 years
industrial works	3 years
public buildings (i.e. cinemas, leisure complexes, restaurants, hotels etc.)	1 year
special installations (i.e. fire alarms, launderettes, petrol filling stations etc.)	1 year

Two Approved Codes of Practice written under the Electricity at Work Regulations deal with electrical systems in mines[23] and quarries[24]. Both these codes require periodic inspection and test of electrical equipment.

In the case of testing class I, IIA, IIB and III portable electrical equipment reference should be made to an HSE Guidance Note[25]. Portable electrical tools should be inspected each time they are returned to stores and tested annually, although under certain hostile operating conditions testing should be more frequent.

The testing of electrical installations and equipment should include:

- polarity,
- earth fault loop impedance/earth continuity,
- insulation resistance,
- operation of devices for isolation and switching,
- operation of residual current devices,
- verification of overcurrent protective devices
- relay and starter contact erosion, and
- wiring integrity.

Portable equipment should be tagged to identify it and indicate its inspection status. This can provide the data for a register which can be used to maintain a record of inspection dates. Similar records should be kept of inspections and examinations of fixed installations.

5.8 Other considerations

A number of common issues arise from the various legislative requirements for examination and inspection of plant and equipment.

5.8.1 Competent person

In general where the law requires particular items of plant to be thoroughly examined and tested it requires that the work be carried out by a 'competent person'.

There is no precise guidance in the statute or in subsequent case law as to what constitutes 'competence'. However, the competency of a person to carry out particular examinations or tests is a matter of fact on which the occupier or owner of the statutory equipment must be satisfied. In the event of legal proceedings it will require to be demonstrated in court that the person chosen was indeed competent to carry out the statutory surveys.

An often quoted definition is that 'the competent person should have such practical and theoretical knowledge and actual experience of the type of machinery or plant which he has to examine, as will enable him to detect defects or weaknesses which it is the purpose of the examination to discover, and to assess their importance in relation to the strength of the machinery or plant in relation to its function'. It is not sufficient for the person making the examination to be able to detect faults, he must also, from his knowledge and experience, be able to assess their seriousness.

The competent person need not necessarily be from an independent authority, but can be an internal appointment of a suitably qualified person by the company concerned, but in exercising his responsibilities for carrying out examinations he must be separated from any other functions that might cause a clash of interests. The competent person, whether employee or from a specialist organisation, must be allowed to act objectively and in a professional manner.

For complex plant and equipment it is doubtful whether any one individual would have sufficient knowledge and expertise to carry out the full examination on his own. In such circumstances, the support of a team of suitably qualified specialists may be needed.

The Code of Practice on Pressure Systems[1] considers that competence should be related to the size, complexity and hazard associated with the plant concerned and suggests different degrees of qualification for minor, intermediate and major pressure systems. The largest employers of competent persons, the engineer-surveyors, are the specialist engineering inspecting authorities associated with insurance companies, but there is an increasing number of small inspection companies providing 'competent person' inspection services who have no interest in insurance. While there is no requirement in law to have any item of plant, whether subject to statutory inspection or not, insured, it is prudent to do so. Some large companies – particularly in the steel and petrochemical industries – have their own in-house inspection departments.

The onus of responsibility for compliance with the statutory inspection obligations has been placed firmly with the company or organisation that owns the plant, and the enforcing authorities have always taken a consistent and firm line on this point. Thus, if a specialist inspection organisation is engaged to carry out all statutory examinations and it fails to do so, either properly or within the statutory period of time, the inspecting organisation may be in breach of contract, but it is the company that owns the plant that is held responsible for the breach of the particular legislation.

However, Regulations recognise the fact that a person, other than the owner and not employed by him, may, by his actions or failure to act, cause a breach resulting in an offence and offer the owner a defence if he identifies the person who caused the breach and can prove that he – the owner – exercised all due diligence to avoid the breach. Nevertheless, under s. 36 of HSW the owner may still be liable.

Under certain regulations there is a requirement for a 'responsible person' to carry out routine daily or weekly inspections of plant. That responsible person could be the employee who regularly operates the plant or equipment provided he has been trained in the checks to be carried out.

In recent years a national voluntary accreditation scheme – based on BS EN 45004[26] – has been established for inspection bodies under the aegis of the United Kingdom Accreditation Service (UKAS). Inspection bodies operating within this scheme are required to establish and maintain a quality system within their scope of activities which is subject to ongoing assessment by UKAS. The scheme is supported by HSE.

5.8.2 Patent and latent defects

When referring to requirements to be met by various lifting and boiler plant, the phrase is used in the FA that 'the machine shall be of good construction, sound material, adequate strength', and sometimes the words 'and free from patent defect' are added. While no definition is given of this latter phrase, its meaning seems to relate to the design of the equipment. Similarly, latent – which means hidden – would seem to refer to the material used to construct the item. The only guidance on this is from *McNeil* v. *Dickson and Mann*[27] where it was held that it was irrelevant whether a defect was patent or latent since the words did not qualify the major requirement 'to be of good construction, sound material and adequate strength'.

In considering whether a machine is of 'good mechanical construction', regard must be given to the stresses imposed upon the item when used for its intended purpose and to any additional stresses that could conceivably be imposed upon the item during normal use, due to impact loading, excessive vibration, inertia forces or temperature (both elevated or sub-zero).

'Sound material' means in fact material which is inherently intact and without internal defect, and not material which merely appears to be sound.

Whether plant is of 'adequate strength' must be related to its ability to withstand the stresses and strains met in its intended or foreseeable use.

5.8.3 Thorough examination

A thorough examination is a detailed visual examination, both stationary and under working conditions, carried out as carefully as the conditions permit in order to arrive at a reliable conclusion as to the safety of the plant or machinery and hence an assurance that the operation of the plant will be safe until the next statutory inspection.

Where the competent person needs to satisfy himself as to the condition of internal component parts he can require the plant to be dismantled. In addition, whenever considered necessary the visual examination can be supplemented by non-destructive testing of components to determine their internal or surface condition without causing any detrimental change in the material.

5.8.4 Reporting

The completion of a report of an examination is a statutory requirement in respect of certain types of plant and it can be made by separate reports or by annotating the register of plant, or a combination of both.

Reports of examinations can be in a form to suit the existing record keeping system although they should contain as a minimum the particulars listed in the ACOP[3]. The reports of an examination usually indicate the condition of the plant, e.g. 'in good order', list the defects requiring correction or place restrictions on the use of the plant because of its condition. Where defects are found, a copy of the report has to be sent to the enforcing authority within a prescribed time limit. Defects can affect the safe working of the plant or machinery and where identified must receive the attention specified within the time limit given in the report.

It is common practice for the engineer-surveyor to make observations in his report to draw attention to other matters of a less serious nature. These observations are advisory and do not affect the continuing safe operation of the plant and machinery.

The reports of the statutory inspections must be kept readily available for inspection by the enforcing authorities. Where needed as working documents, photocopies of the report should be used so that the original can be kept as a master copy in a central file.

5.9 Conclusion

This chapter has endeavoured to cover the principal statutory inspection requirements in the UK that are likely to be of concern to occupational safety advisers. Certain areas have not been covered, such as gasholders and builders' hoists nor the slightly differing requirements in Eire, Northern Ireland, the Channel Islands and the Isle of Man.

It should not be assumed that because there is no statutory requirement to periodically inspect a particular type of plant or machine, that it need not be so inspected. Section 2 of HSW places a duty on employers to provide and maintain plant in a safe condition and to provide a safe place to work generally. To meet this general obligation it is prudent to carry out regular inspections and tests on a range of plant and machines and to record the results. This is underlined by the increasing number of industry-produced Codes of Practice, such as for injection moulding machines, die-casting machines and concrete pumping booms, which recommend inspections and tests as being the best practical means of ensuring continuing compliance with HSW.

Modern legislation permits inspection and test requirements to be determined following a risk assessment conducted by the user/owner. This gets away from the prescriptive requirements of earlier legislation which had proved inappropriate in a number of circumstances. The current goal setting approach ensures the user/owner has much greater flexibility in selecting appropriate inspection regimes but at the same time imposes a responsibility on him to select suitable and effective regimes which ensure the continuing safety of work equipment.

References

1. Health and Safety Executive, Approved Code of Practice No. COP 37, *Safety of pressure systems. Pressure Systems and Transportable Gas Containers Regulations 1989*, HSE Books, Sudbury (1990)
2. Health and Safety Executive, Approved Code of Practice No. COP 38, *Safety of transportable gas containers. Pressure Systems and Transportable Gas Containers Regulations 1989*, HSE Books, Sudbury (1990)

3. Safety Assessment Federation, *Guidance on the examination of longitudinal seams*, SAFed, London (to be published)
4. Safety Assessment Federation, *Guidelines on periodicity of examinations*, SAFed, London (1997)
5. Safety Assessment Federation, *Guidelines on the examination of boiler shell to end plate and furnace to end plate welded joints*, SAFed, London (1997)
6. Health and Safety Executive, Guidance Note No. PM 60, *Steam boiler blowdown systems*, HSE Books, Sudbury (1987)
7. Milton, J.H. and Leach, R.M., *Marine Steam Boilers*, 4th edn, Butterworth-Heinemann, Oxford (1980)
8. EU, *Amending Directive to the Use of Work Equipment Directive No. 95/63/EC*, EU, Luxembourg (1995)
9. Health and Safety Executive, Guidance Note No. PM 8, *Passenger carrying paternosters*, HSE Books, Sudbury (1977)
10. Health and Safety Executive, *The Lifting Plant and Equipment (Records of Test and Examination etc.) Regulations 1992*, HMSO, London (1992)
11. Health and Safety Executive, Legislation Booklet L20, *A guide to the Lifting Plant and Equipment (Records of Test and Examination etc.) Regulations 1992*, HSE Books, Sudbury (1992)
12. Health and Safety Executive, Guidance Note No. PM 7, *Lifts: thorough examination and testing*, HSE Books, Sudbury (1982)
13. British Standards Institution, proposed *BS EN 81, Safety rules for the construction and installation of lifts*, BSI, London (to be published)
14. Health and Safety Executive, Guidance Note No. PM 79, *Power presses: Thorough examination and testing*, HSE Books, Sudbury (1995)
15. Health and Safety Executive, *Press brakes*, HSE Books, Sudbury (1984)
16. Health and Safety Executive, Guidance Booklets Nos: (a) HS(G) 37, *Introduction to local exhaust ventilation* (1993) (b) HS(G) 54, *The maintenance, examination and testing of local exhaust ventilation* (1990). HSE Books, Sudbury
17. Health and Safety Executive, Legislation Booklet No. L 5, *General COSHH ACOP and Carcinogens ACOP and Biological Agents ACOP (1996 edition)*, HSE Books, Sudbury (1997)
18. Health and Safety Executive, Approved Code of Practice No. COP 30, *Control of substances hazardous to health in fumigation operations*, HSE Books, Sudbury (1988)
19. Health and Safety Executive, Legislation Booklet No. L27, *The control of asbestos at work*, HSE Books, Sudbury (1993)
20. Health and Safety Executive, Approved Code of Practice No. 2, *Control of lead at work*, HSE Books, Sudbury (1998)
21. The Institution of Electrical Engineers, *Regulations for Electrical Installations*, 16th edn, IEE, London (1991)
22. British Standards Institution, *BS 7671 Requirements for electrical installations*, BSI, London (1989)
23. Health and Safety Executive, Approved Code of Practice No. COP 34, *The use of electricity in mines*, HSE Books, Sudbury (1989)
24. Heath and Safety Executive, Approved Code of Practice No. COP 35, *The use of electricity in quarries*, HSE Books, Sudbury (1989)
25. Health and Safety Executive, Guidance Note No. PM 32, *Safe use of portable electrical apparatus (electrical safety)*, HSE Books, Sudbury (1990)
26. British Standards Institution, *BS EN 45004 General criteria for the operation of various types of bodies performing inspections*, BSI, London
27. McNeil *v.* Dickson and Mann (1957) SC 345

Further reading and references

General

Sinclair, T. Craig, *A Cost-Effective Approach to Industrial Safety*, HMSO, London (1972)

Legal

Fife, I. and Machin, E.A., *Redgrave Fife and Machin; Health and Safety*, Butterworth-Heinemann, Oxford (1993)

Munkman, J., *Employer's Liability at Common Law*, 11th edn, Butterworth, London (1990)

Pressure vessels
Jackson, J., *Steam Boiler Operation: Principles and Practices*, 2nd edn, Prentice-Hall, London (1987)
Robertson, W.S., *Boiler Efficiency and Safety*, MacMillan Press, London
Brown, Nickels and Warwick, *Periodic Inspection of Pressure Vessels*, (A.O.T.C.) I. Mech. E. Conference, London (1972)
British Standards Institution, London:
BS 470:1984 Specification for inspection, access and entry openings for pressure vessels
BS 709:1983 Methods of destructive testing fusion welded joints and weld metal in steel
BS 759 (Pt 1):1984 Specification for valves, gauges and other safety fittings for application to boilers, and to piping installations for and in connection with boilers
BS 1113:1992 Specification for design and manufacture of water tube steam generating plant (including superheaters, reheaters and steel tube economisers)
BS 1123:1987 Specification for safety valves, gauges and other safety fittings for air receivers and compressed air installations
BS 2790:1992 Specification for the design and manufacture of shell boilers of welded construction
BS 2910:1986 Methods for radiographic examination of fusion welded circumferential butt joints in steel pipes
BS 5169:1992 Specification for fusion welded steel air receivers
BS 5500:1997 Specification for unfired fusion welded pressure vessels
BS 6244:1982 Code of Practice for stationary air compressors

ANSI/ASME Boiler and pressure vessel code
Sec. I – Power boilers
Sec. VII – Recommended guidelines for the care of power boilers

HSE Guidance Notes, HSE Books, Sudbury
GS 4 Safety in pressure testing (1992)
GS 5 Entry into confined spaces (1995)
GS 20 Fire precautions in pressurised workings (1983)
PM 5 Automatically controlled steam and hot water boilers (1989)
HS (G) 29 Locomotive boilers (1986)
HS(R)30 *A guide to the Pressure Systems and Transportable Gas Containers Regulations 1989*
PM 60 Steam boiler blowdown systems (1987)

Lifting and handling plant
Phillips, R.S., *Electric Lifts*, Pitman, London (1973)
Dickie, D.E., *Lifting Tackle Manual*, (Ed. Douglas Short), Butterworth, London (1981)
Dickie, D.E., *Crane Handbook*, (Ed. Douglas Short), Butterworth, London (1981)
Dickie, D.E., *Rigging Manual*, Construction Safety Association of Ontario (1975)
Associated Offices Technical Committee (A.O.T.C.) *Guide to the testing of cranes and other lifting machines*, 2nd edn, A.O.T.C., Manchester (1983)
British Standards Institution, London:
BS 466:1984 Specification for power driven overhead travelling cranes, semi-goliath and goliath cranes for general use
BS 1757:1986 Specification for power driven mobile cranes
BS 2452:1954 Specification for electrically driven jib cranes mounted on a high pedestal or portal carriage (high pedestal or portal jib cranes)
BS 2573 (Pt 1): 1983 Specification for the classification, stress calculations and design criteria for structures
BS 2573 (Pt 2): 1980 Specification for the classification, stress calculations and design of mechanisms
BS 2853: 1957 Specification for the design and testing of steel overhead runway beams
BS 4465: 1989 Specification for design and construction of electric hoists for both passengers and materials
BS 5655 (10 parts) covering safety of electric and hydraulic lifts, dimensions, selection and installation, control devices and indicators, suspension eyebolts, guides, and the testing and inspection

BS 5744: 1979 Code of Practice for the safe use of cranes
ISO 4309 Wire ropes for lifting appliances – Code of Practice for examination and discard
ISO 4310 Cranes – test code and procedures
HSE Guidance Notes, HSE Books, Sudbury
PM 3 Erection and dismantling of tower cranes (1976)
PM 7 Lifts: thorough examination and testing (1982)
PM 8 Passenger carrying paternosters (1987)
PM 9 Access to tower cranes (1979)
PM 24 Safety at rack and pinion hoists (1981)
PM 26 Safety at lift landings (1981)
PM 27 Construction hoists (1981)
PM 34 Safety in the use of escalators (1983)
PM 43 Scotch derrick cranes (1984)
PM 45 Escalators: periodic thorough examination (1984)
PM 54 Lifting gear standards (1985)
PM 55 Safe working with overhead travelling cranes (1985)
PM 63 Inclined hoists used in building and construction work (1987)
HS(G) 19 Safety in working with power operated mobile work platforms (1982)
HS(G) 23 Safety at power operated mast work platforms (1985)
L20 *A guide to the Lifting Plant and Equipment (Records of Test and Examination etc.) Regulations 1992*

Power presses
Joint Standing Committee on Safety in the Use of Power Presses:
Safety in the use of power presses, HSE Books, Sudbury (1979)
Power press safety; Safety in material feeding and component ejection systems, HSE Books, Sudbury (1984)
British Standards Institution, London:
BS 4656 (Pt 34): 1985 Specification for power presses, mechanical, open front
BS 6491: Pt 1:1984 Electro sensitive safety systems for industrial machines
Pt 2:1986 Particular requirements for photo-electric sensing units
[Note: to be replaced by: BS EN 50100, Electro-sensitive protective systems,
part 1 General requirements (to be published)
part 2 Opto-electronic devices (to be published)]

HSE Guidance Notes, HSE Books, Sudbury
PM 41 Application of photo-electric safety systems to machinery (1984)

Local exhaust ventilation plant
Industrial Ventilation, American Conference of Government Industrial Hygienists, Cincinatti, Ohio

Principles of Local Exhaust Ventilation, Report of the Dust and Fume Sub Committee of the Joint Standing Committee on Health, Safety and Welfare in Foundries, HSE Books, Sudbury (1975)
Relevant Standard:
BS 6540 (Pt 1):1985 Methods of test for atmospheric dust spot and efficiency, and synthetic dust weight arrestance
HSE Guidance Notes, HSE Books, Sudbury
MS 13 Asbestos (1988)
EH 10 Asbestos – Exposure limits and measurements of airborne dust concentrations (1995)
EH 25 Cotton dust sampling (1980)
EH 28 Control of lead: air sampling techniques and strategies (1986)
EH 30 Control of lead: pottery and related industries (1981)
HS(G) 37 Introduction to local exhaust ventilation (1993)

Electrical installations
Institution of Electrical Engineers, *Regulations for Electrical Installations*, 16th edn, London (1991)

British Standards Institution, London:
BS 2754:1976 Construction of electrical equipment for protection against electric shock
BS EN 60204, Safety of machinery – Electrical equipment of machines, part 1 Specification for general requirements
BS 4444:1989 Guide to electrical earth monitoring
BS 5958 (Pt 1):1991 Code of Practice for control of undesirable static electricity – general considerations
BS 5958 (Pt 2):1991 Code of Practice for control of undesirable static electricity – recommendations for particular industrial situations
BS 6233:1982 Methods of test for volume resistivity and surface resistivity of solid electrical insulating materials
HSE Guidance Notes, HSE Books, Sudbury
GS 6 Avoidance of danger from overhead electrical lines (1997)
PM 29 Electrical hazards from steam/water pressure cleaners etc. (1995)
PM 32 The safe use of portable electrical apparatus (1996)
PM 38 Selection and use of electric hand lamps (1991)

Chapter 6

Safety on construction sites

R. Hudson

The construction industry has always been plagued with an abundance of reportable accidents coming to an all time high of over 45 000 accidents in 1966 with, over the previous decade, an average of 250 persons killed each year. These appalling figures occurred in spite of a considerable volume of safety legislation aimed at improving safe working in the construction industry.

Many of the causes of these accidents are reflected in the detailed requirements of the relevant Regulations[1] which lay down the preventive measures to be taken. This chapter looks at the safety legislation for the construction industry and some of the techniques for meeting the required safety standards.

It should be remembered, however, that since the Health and Safety at Work etc. Act 1974 (HSW) came into effect all subordinate legislation, such as Regulations, made under it apply to all employer/employee relationships and though the title may not include 'construction' this does not mean they do not apply to construction works.

6.1 Construction accidents

An indication of the size and seriousness of the problem can be obtained by considering the annual HSE report[2] containing data on fatal and major accidents and respective incidence rates.

While overall until 1996–7 there has been a reduction in fatal accidents this may be accounted for by a smaller workforce or the change in, rather than improved, standards. If the incidence rate for construction projects, covering the range from large civil and high rise building to refurbishment and low rise structures, is compared with manufacturing it is, year on year, consistently six times more dangerous. As some 70% of the accidents investigated could have been prevented by management action[3] this continued to be an unacceptable situation.

Further analysis of both fatal and major accidents gives a good indication of the problem areas. While the numbers vary from year to year the pattern remains fairly constant with 'falls from height' accounting for some 40% of major injuries and 50% of fatalities.

6.2 Safe working in the industry

In considering safe working and accident prevention in the construction industry, this chapter will follow broadly the progression of a construction operation. All stages should be adequately planned making allowance for the incorporation of safe systems of work.

Planning has been the province of the main contractor but with the coming into effect of the Construction (Design and Management) Regulations 1994[4] (CDM) this responsibility has been clarified. Under these Regulations the client has an obligation to appoint a competent planning supervisor for the project. In many instances this role will be filled by professional advisers such as architects or engineers who act on behalf of the client.

The planning supervisor is required to:

- ensure the designers have fulfilled their responsibilities under the regulations and the design includes adequate information about the design and materials to be used where they might affect the health and safety of those carrying out the construction work;
- prepare a health and safety plan, to be included with the tender documentation, which details the risks to health and safety of any person carrying out the construction work so far as is known to the planning supervisor or are reasonably foreseeable, and any other relevant information to enable the contractor to manage the works;
- prepare and deliver to the client a health and safety file on the as-built structure which the client retains for reference during subsequent construction works on the structure.

The client is required to appoint a competent principal contractor for the project.

The principal contractor must for his part:

- adopt and develop the health and safety plan and provide information for the health and safety file;
- ensure the health and safety plan is followed by all persons on the site; and
- co-ordinate the activities of others on the site and ensure that all co-operate in complying with the relevant statutory provisions that affect the works.

For these purposes the principal contractor can give directions or establish rules for the management of the construction works as part of the health and safety plan. Such rules must be in writing and be brought to the attention of all affected persons.

Finally, one of the main provisions of these Regulations defines the responsibility which designers, such as architects, have for health and safety during the construction stages. Designers have to ensure, so far as is reasonably practicable and provided the structure conforms to their design, that persons building, maintaining, repairing, repainting, redecorating or cleaning the structure are not exposed to risks to their health and safety. In addition, the designer must ensure that included in the design documentation is adequate information about the design and materials used, particularly where they may affect the health and safety of persons working on the structure.

Basically, these requirements place designers of buildings and structures, such as architects, under similar obligations to those who design articles and substances, whose obligations are contained in s.6 of HSW.

As work gets under way, the principal contractor, who has responsibility for the construction phase of the project, has to ensure that all those employed are properly trained for their jobs. Under HSW, now amplified by the Management of Health and Safety at Work Regulations 1992[1] (MHSWR), the employer is required to provide training in specific circumstances, i.e. on joining an employer, when work situations change and at regular intervals.

In addition, specific job training is prescribed in numerous statutory provisions such as the Abrasive Wheels Regulations 1970 (for mounting abrasive wheels) and the Provision and Use of Work Equipment Regulations 1992[1] (PUWER) (adequate training in the use of work equipment).

In an industry increasingly reliant upon the use of subcontractors, the main contractor retains the onus for health and safety on site. This onus can extend to training employees of subcontractors where their activities may affect the health and safety of the employees of the main contractor and of the subcontractor himself (ss. 2 and 3 of HSW). This is more clearly defined by CDM which requires the principal contractor to ensure that other employers on the construction work provide their employees with appropriate health and safety training when they are exposed to new or additional risks due to:

- changes of responsibilities, i.e. promotion,
- use of new or changed work equipment,
- new technology, or
- new or changed systems of work.

The provision of information is also an essential contribution to reducing risks to health and safety. As with training, ss. 2 and 3 of HSW apply to both main and subcontract employers in relation to informing each other of risks, within their knowledge, arising out of their work. The decision in *Regina* v. *Swan Hunter Shipbuilders Ltd*[5] clarified this in respect of 'special risks'. In this case, a number of fatalities resulted from a fire in a poorly ventilated space in a ship which had become enriched with oxygen due to an oxygen supply valve being left open by a subcontractor's employee. An intense fire developed when another contractor struck an electric arc to do some welding. Swan Hunter were well aware of the fire risk associated with oxygen enrichment and provided detailed information for their own, but not subcontractors' employees. Swan Hunter were prosecuted under ss. 2 and 3 of HSW and convicted for failing to ensure the health and safety of their own employees by not informing the employees of subcontractors of special risks which were within its, Swan Hunter's, knowledge, i.e. from fires in oxygen enriched atmospheres. This decision has been overtaken by MHSWR (reg. 9) which requires employers who share a workplace to take all reasonable steps to inform other employers of any risks arising from their work. Further, CDM places a duty on the principal contractor to inform other contractors of the risks arising out of or in connection with the works and ensure that those subcontractors inform their employees of:

- risks identified by the contractor's own general risk assessment,
- the preventive and precautionary measures that have to be implemented,
- any serious or imminently dangerous procedures and the identity of any persons nominated to implement those procedures, and
- details of the risks notified to him by the principal or another contractor.

Apart from the overall obligations placed on both the main and sub-contractor employers by the Health and Safety at Work Act, more extensive requirements specific to the building and construction industry are contained in Regulations dealing with particular aspects of safety in building and construction work.

6.2.1 Notification of construction work

The CDM Regulations have redefined the work that has to be notified extending it from 'building operation or work of engineering construction' to a broader term 'construction work'. This latter term is defined in the Regulations as including every aspect of the carrying out of the work from beginning to end of a project. It includes site clearance and site investigation, the assembly and disassembly of fabricated units (site huts), the demolition and removal of spoil and the installation, commissioning, maintenance and repair of services such as telephones, electricity, compressed air, gas etc. The interpretation of 'construction work' is being extended to include the engineering work involved in the installation, maintenance and dismantling of major process plants.

Responsibility for making the notification lies with the planning supervisor who must provide to the HSE the information listed in schedule 1 of the Regulations before any work starts on the site. Official form F10 (rev), which calls for the necessary information, can be used but is not mandatory. Notification of construction work must be made where the work being undertaken is expected to last more than 30 days or where it involves a total of more than 500 person-days. A working day is any day on which work of any sort is carried out on the site and includes weekends and all other times outside the 'normal' working week.

However, whether there is a need to notify or not, full compliance with all the requirements of the relevant health and safety legislation is necessary.

6.2.2 The Construction (Health Safety and Welfare) Regulations 1997

6.2.2.1 Responsibilities

The responsibility for complying with the requirements of these Regulations is placed on the self-employed person, the person controlling construction works, employees and every person at work.

The requirements of the Regulations cover several subject areas which are dealt with in greater depth below, by including practical advice on the separate subjects to give a greater understanding of how compliance with the Regulations can be achieved.

6.2.2.2 Safety in excavations

In any excavation, earth work, trench, well, shaft, tunnel or underground working where there is a risk of material collapsing or falling, proper support must be used as early as practicable in the course of the work to prevent any danger from an earth fall or collapse. Suitable and sufficient material should be available for this purpose or alternative methods used such as:

1 *Battering the sides*, i.e. cutting the sides of the excavation back from the vertical to such a degree that fall of earth is prevented.
2 *Benching the sides*. The sides of the excavation are stepped to restrict the fall of earth to small amounts. Maximum step depth 1.2 m (4 ft).

Figure 6.1 shows typical examples of these trenching techniques.

Inspection of any excavation which is supported must be made:

(a) at the start of every shift before any person carries out any work;
(b) after any event likely to have affected the strength or stability of the excavation or any part of it;
(c) after any accidental fall of rock, earth or other material.

A report of the inspection containing the prescribed particulars (no form or official register is necessary) shall be made within 24 hours of the inspection and retained until 3 months after the work has been completed. Only one report needs to be made every 7 days in respect of item (a) above for excavations and items (a) and (b) for coffer-dams and caissons. Reports of inspections following the other incidents listed above must be made before the end of the working period.

All material used for support should be inspected before use and material found defective must not be used. Supports must only be erected, altered or dismantled under competent supervision and when-

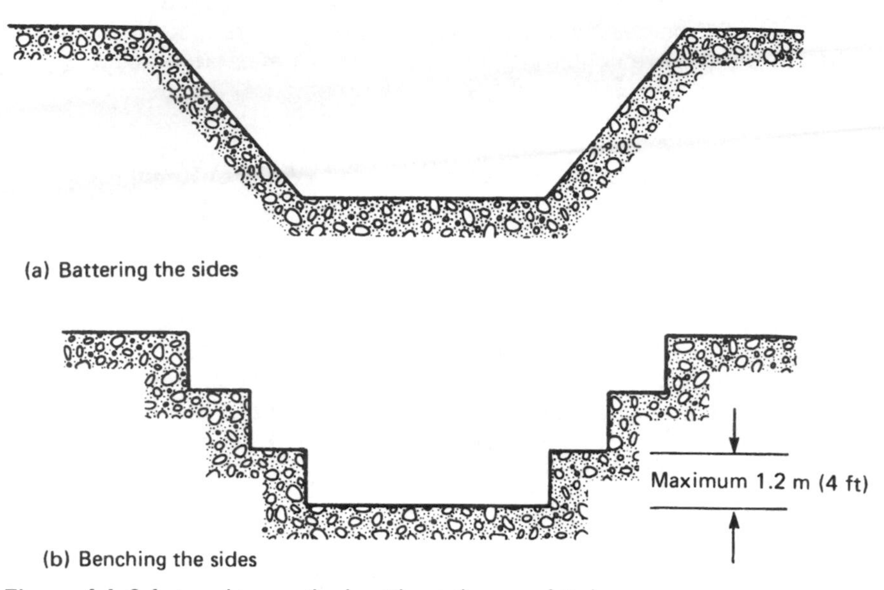

Figure 6.1 Safe trenching methods without the use of timber

ever practicable by experienced operatives. All support must be properly constructed and maintained in good order. Struts and braces must be fixed so that they cannot be accidentally dislodged. In addition, in the case of a coffer-dam or caisson, all materials must be examined and only if found suitable should they be used.

If there is risk of flooding, ladders or other means of escape must be provided.

When excavating in close proximity to existing buildings or structures, be they permanent or temporary, there is a requirement to give full consideration to their continued stability. This is intended to protect persons employed on site. However, under the Health and Safety at Work Act this responsibility is extended to the safety of the public, i.e. those not employed on the site, and may relate to private dwelling houses, public buildings or public rights of way. It is particularly important when excavating near scaffolding.

Where any existing building or structure is likely to be affected by excavation work in the vicinity, shoring or other support must be provided to prevent collapse of the building or structure. Examples of trench shoring are given in *Figures 6.2, 6.3* and *6.4*.

Excavations more than 2 m deep near which men work or pass, must be protected at the edge by guardrails or barriers or must be securely covered. Guardrails, barriers or covers may be temporarily moved for access or for movement of plant or materials but must be replaced as quickly as possible.

Where the excavation is not in an enclosed site and is accessible to the public the standard for protection is more onerous. Even the most

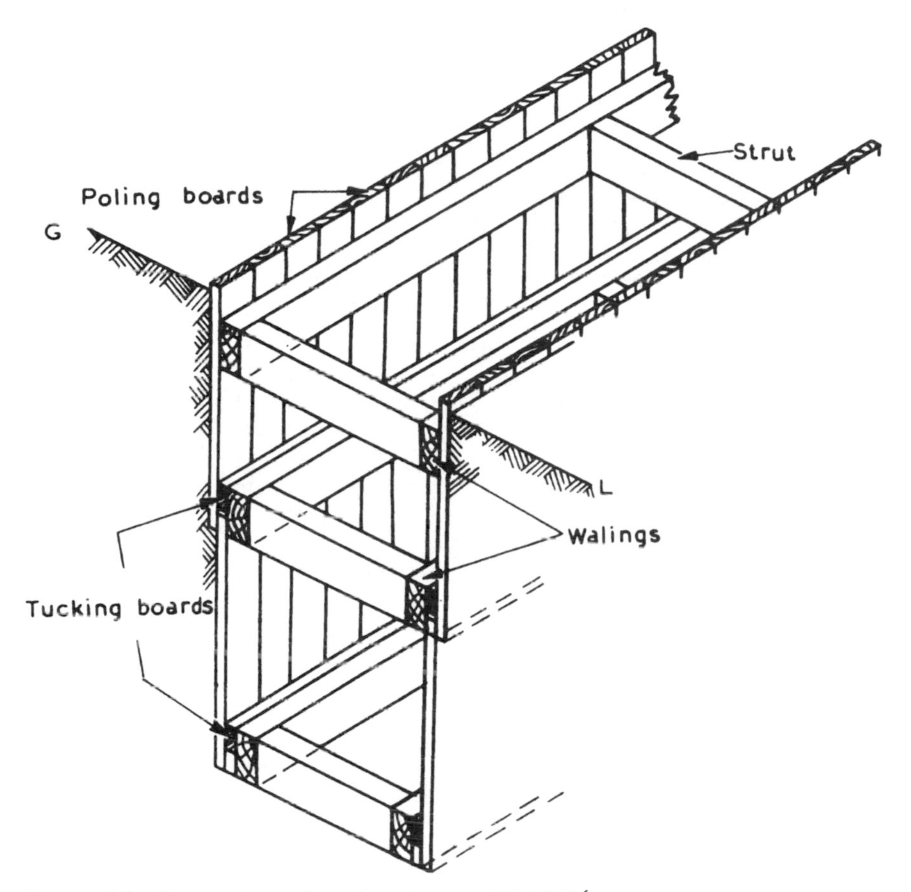

Figure 6.2 Close poling with tucking frames. (BS 6031)[6]

shallow depressions should be fenced so that members of the public are not exposed to risks to their health and safety.

Materials, plant, machinery etc. must be kept away from the edge of all excavations to avoid collapse of the sides and the risk of men falling in, or material falling on men.

Severe weather conditions such as heavy rain, or where timber has become wet then followed by a hot dry spell, could so affect timbering etc. as to cause it to become dangerous. In these circumstances, where the strength or stability of an excavation could be adversely affected, an inspection would be required together with a report. Guidance on the construction of trenches, pits and shafts is given in the British Standard CP 6031 Code of Practice for Earthworks[6].

On sites where mobile machinery such as tippers, diggers, rough terrain fork lift trucks etc. are used, special care should be taken to ensure that operators are fully aware of the stability of their machines and of the maximum slope on which they can be safely used. Particular attention

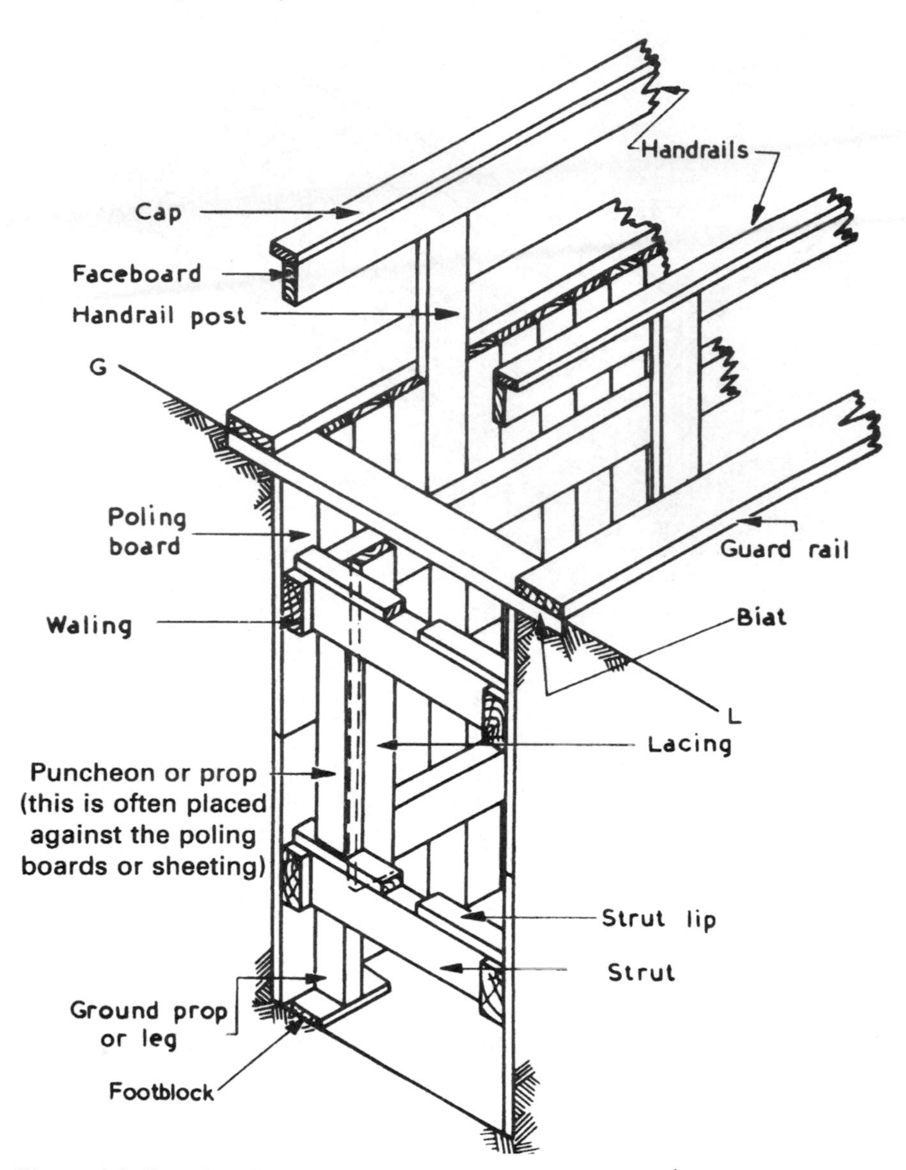

Figure 6.3 Typical single or centre waling poling frame. (BS 6031[6])

should be paid to the condition of the ground and whether it is capable of bearing the vehicle weight. Information on safe ground conditions and angles of tilt can be obtained from the machine manufacturers.

Where overhead cables cross the line of excavations, particular care must be taken in the selection of the type of plant to be used and precautions taken to ensure that the equipment does not or cannot touch live high voltage conductors. Underground cables, be they high or low voltage, telephone or television links, together with gas piping, present a

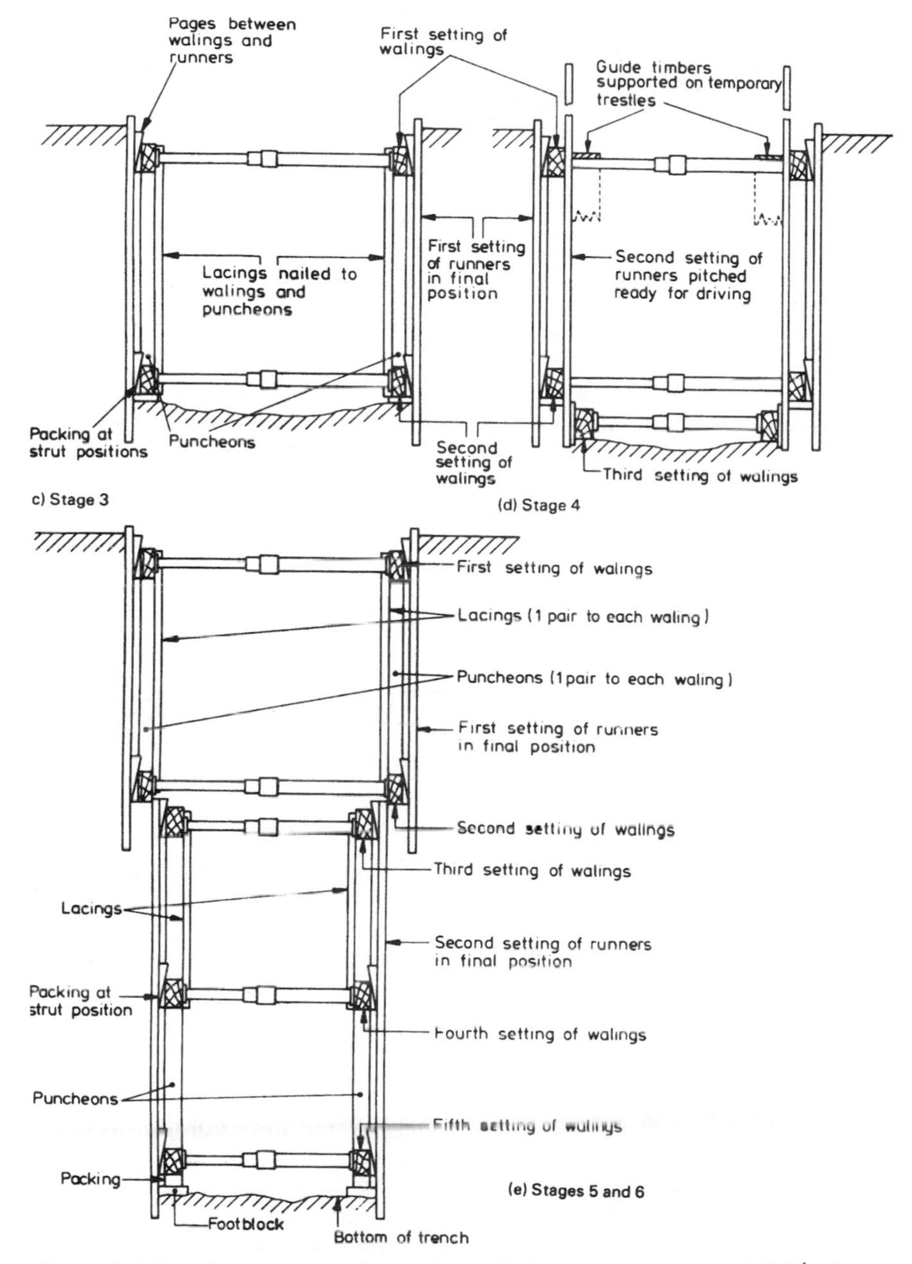

Figure 6.4 Trench excavation using steel trench sheets as runners. (BS 6031)[6]. Note Stages 1 and 2 are shown in the British Standard

more difficult problem which, in the main, rests with the excavating contractor. Advice is given in an HSE publication[7] but the contractor should approach each of the service authorities asking for accurate information on the actual location, run and depth of their services. This should be in writing, and the information supplied by the authority should preferably be marked on a drawing; ideally the authority should authorise the drawing as correct. The location should be confirmed using devices for locating cables and other services and the route of the service marked on the surface. Services should be carefully exposed by hand-dig methods to verify their precise location and depth before mechanical means are employed. Mechanical equipment such as excavators should not be used within 0.5 m of the suspected cable until its route has been specifically located.

Knowledge of the whereabouts of underground services is also necessary where heavy plant or vehicles are used since many such services are at shallow depth and can be damaged by the sheer weight of equipment. Although injury is not likely to result, considerable cost to the contractor could be involved.

Careful planning, including the selection of the correct plant and equipment is essential, for both safety and economic reasons, when carrying out excavation work. For example, the correct size of excavator can act as a crane eliminating the need to bring extra specialised plant onto site. However, when using an excavator in this manner the machine must comply with either all the requirements of the Construction (Lifting Operations) Regulations 1961 that apply to cranes or to the less onerous but restrictive Certificate of Exemption[8] which exempts an excavator from certain requirements of the Regulations applicable to cranes but imposes alternative conditions. However, under the Exemption the excavator is limited to operations immediately associated with excavation work, i.e. lifting pipes into an excavation. It may not be used for lifting items from the back of a lorry into a storage compound or subsequently moving them to an excavation. However, this complex situation may not continue for much longer since the Construction (Lifting Operations) Regulations 1961 are due to be replaced by legislation having a risk-based approach to the inspection, examination and testing of lifting equipment.

Before work is started on a construction site there are a number of matters that should be checked, from both a prudent and statutory point of view:

1 Provide site security – particularly to stop children getting in.[9]
2 Investigate the nature of the ground before excavations begin and decide the form that the support work will need to take and ensure that adequate supplies of sufficiently strong materials are available. Special precautions may be needed where trenches pass near adjacent roads or buildings.
3 Locate all public services, water, gas, electricity, telephone, sewers etc., and avoid if possible; if not, take necessary precautions.
4 Provide material for barriers and authorise traffic notices.
5 Provide adequate lighting.

6 Position spoil heap at a distance not less than the depth from the edge of excavation. Tip if possible on blind side of excavator to ensure operator has visibility when swinging back to trench excavation.
7 Provide personal protective equipment.
8 Provide sufficient ladders of suitable length, strength and type.
9 Query necessity for bridges and gangways.
10 Take note of all overhead services, the arrangements made for their protection and the safety of all working in their vicinity.

6.2.2.3 Mechanical plant and portable tools

All machinery for use at work is now subject to the Provision and Use of Work Equipment Regulations 1992[1] (PUWER) with its requirements for guarding dangerous parts, controls and associated matters.

The general principles of guarding are contained in BS 5304[10] but this standard is being replaced by a number of harmonised European (EN) standards. Other publications[11] are available that give advice on the safety and care of site plant and equipment.

Where guards are removed to enable maintenance work to be carried out, they must be replaced before the machinery is returned to normal work.

Portable tools are used extensively on construction sites and commonly suffer damage. Arrangement should be made for regular checks on the condition of portable tools, paying particular attention to the integrity of electrical insulation on the tool itself and on the lead and to damage to rotating parts remembering that the Electricity at Work Regulations 1989 require such equipment to be properly maintained, involving regular inspection and testing. In the case of compressed air equipment, the Pressure Systems and Transportable Gas Container Regulations 1989 also have effect.

Under PUWER, not only the operators of any plant and machinery, being work equipment, but their supervisors and managers must be trained:

- in the correct method of use,
- on the risks such use may involve, and
- on the precautions to be taken.

In the case of cranes, capstans, winches, woodworking machines and mechanically propelled vehicles, there are legal obligations for them to be so. Operators of these machines must be over 18 years old unless being trained and under the direct supervision of a competent person.

6.3 Site hazards

The extensive use of temporary or semi-permanent wiring on construction sites, the rough usage that equipment gets, the hostile conditions under which it is used and, often, the lack of knowledge of those using the equipment contribute to the high risk potential of the use of electricity. Compliance with the Electricity at Work Regulations[12] will reduce the hazards, which broadly can be divided into three categories:

1 Electrocution.
2 Fire.
3 Glare.

6.3.1 Electrocution

There are three operations that carry the highest risk of electrocution: use of portable tools, striking a buried cable, and cranes and diggers making contact with overhead power lines.

Portable tools are used extensively on sites and maintaining them and their connecting cables in good repair is a critical factor of their safe use. Electrocution occurs when the body acts as the conductor between a power line and earth, often because the earth connection on the tool has broken or, less commonly, as a link between differently charged conductors. All portable tools must be securely earthed or be of double or all insulated construction and the plug on the lead must be correctly fused. Unfortunately it is frequently difficult to keep track of every item, so reliance has to be placed on the person using them.

Protection for 220 V supplies can be obtained by the use of residual current devices (RCDs) either in the supply circuit or on the connection to the particular equipment. In addition, on construction sites 220 V supplies should be carried by armoured, metal sheathed or other suitably protected cables.

The greatest protection from electrical shock on construction sites is through the use of a reduced voltage system. Where portable hand-held tools are used this is essential and the system recommended is 110 V ac with the centre point of the secondary winding tapped to earth, so that the maximum voltage of the supply above earth will be 55 V which normally is reckoned to be non-fatal. A further alternative is to use a low voltage supply at 24 V, but in this case, because of the low voltage, the equipment tends to be heavier than with higher voltages. The risk of electrocution from power tools is increased if they are used in wet conditions.

Electrocution from striking an underground cable can be spectacular when it occurs and the most effective precaution is to obtain clearance from the local Electricity Board or the factory electrical engineer that the ground is clear of cables. If doubt exists, locating devices are available that enable underground cables to be traced.

All too often overhead power cables cross construction sites and are a potential hazard for cranes and mechanical equipment. If the supply cannot be cut off, suspended warning barriers should be positioned on each side and below the level of the cable and drivers warned that they may only pass under with lowered jib.

6.3.2 Fire

Usually caused through overloading a circuit, frequently because of wrong fusing. Repeated rupturing of the fuse should be investigated to find the cause rather than replacing the blown fuse by a larger one in the hope that it will not blow. A second cause can be through water getting into contact with live apparatus and causing a short circuit which results in overheating of one part of the system. Where electrical heaters are used on sites, they should be of the non-radiant type, i.e. tubular, fan or convector heaters. Multi-bank tubular heaters used for drying clothes should be protected to prevent clothing, paper etc. being placed directly on the tubes. This protection can be achieved by enclosing the heaters in a timber frame covered with wire mesh.

6.3.3 Glare

Not usually recognised as a hazard, glare can prevent a crane driver from seeing clearly what is happening to his load and it can cause patches of darkness in accessways that prevent operators from seeing the floor or obstacles. Electric arc welding flash can cause a painful condition known as 'arc-eyes', so welding operations should be shielded by suitable flame-resistant screens. Floodlights are designed to operate at a height of 6 m or more and must never be taken down to use as local lighting as the glare from such misuse could create areas of black shadow and may even cause eye injury. Floodlights should not be directed upwards since they can dazzle tower crane drivers.

6.3.4 Dangerous and unhealthy atmospheres

Conditions under which work is carried out on construction sites is largely dictated by the weather, ranging from soaking wet to hot, dry and dusty and suitable protection for the health of the operators has to be provided. However, there is also a considerable range of substances[13] and working techniques now in use that have created their own hazards. A number of these are considered below.

6.3.4.1 Cold and wet

Cold is most damaging to health when it is associated with wet, as it is then very difficult to maintain normal body temperature. Being cold and wet frequently and for substantial periods may increase the likelihood of bronchitis and arthritis and other degenerating ailments. The effects of cold and wet on the employee's health and welfare can be mitigated by three factors: food, clothing and shelter. Where practicable, shelter from the worst of the wind and wet should be provided by sheeting or screens. The accommodation which has to be provided 'during interruption of work owing to bad weather' could also be used for warming-up and drying-out breaks whenever men have become cold, wet and uncomfortable.

6.3.4.2 Heat

Excessive heat has tended to be discounted as a problem on construction sites in the UK, and cases of heat exhaustion which do occur during heat-waves are often attributed to some other quite irrelevant cause. Common forms of heat stress produce such symptoms as lassitude, headache, giddiness, fainting and muscular cramp. Sweating results in loss of fluids and salt from the body, and danger arises when this is not compensated for by increased intake of salt and fluids. If the body becomes seriously depleted it can lead to severe muscular cramps.

6.3.4.3 Dust and fumes

Despite the general outdoor nature of the work, construction workers are not immune from the hazards of airborne contaminants. Although natural wind movement will dilute dust and fumes throughout the site, operatives engaged on particular processes may have a dangerous concentration in their immediate breathing zones unless suitable extraction is provided. This is particularly relevant for work in shafts, tunnels and other confined spaces where forced draught ventilation may have to be provided.

Certain processes commonly met on construction sites create hazardous dusts and fumes. Typical are:

Cadmium poisoning from dust and fumes arising from welding, brazing, soldering or heating cadmium plated steel.

Lead poisoning resulting from inhalation of lead fumes when cutting or burning structures or timber that has been protected by lead paint.

Silicosis due to inhaling siliceous dust generated in the cleaning of stone structures, polishing and grinding granite or terrazo.

Carbon monoxide poisoning caused by incomplete combustion in a confined space or from the exhausts of diesel and petrol engines.

Metal fume fever from breathing zinc fumes when welding or burning galvanised steel.

Each of these hazards would be eliminated by the provision of suitable and adequate exhaust ventilation or, in the case of silicosis, by the provision of suitable breathing masks. In each case food should not be consumed in the area, and the medical conditions can be exacerbated through habitual smoking. A good standard of personal hygiene is also an important factor in maintaining good health on site.

6.3.4.4 Industrial dermatitis

The use of an increasing range of chemical based products on sites poses a potential health risk to those who handle them unless suitable precautions are taken. The complaint is neither infectious nor contagious, but once it develops the sufferer can become sensitised (allergic) to the particular chemical and will react to even the smallest exposure. All chemical substances supplied to sites should carry instructions for use on the label and if the precautions recommended by the maker are followed little ill-effect should be experienced.

Barrier creams may be helpful but suffer the disadvantage of wearing off with rough usage or being washed off by water. Effective protection is provided by the use of industrial gloves and, where necessary, aprons, face masks etc. Again good personal hygiene is important and the use of skin conditioning creams after washing is beneficial.

6.3.4.5 Sewers

Sewers, manholes and soakaways are all confined spaces and before any work is carried out in them an assessment of the risks to health and safety from the work to be done must be made to determine the control measures necessary to avoid those risks. Some precautions that may need to be taken include the testing of the atmosphere for toxic and flammable gases and lack of oxygen. Where the atmosphere is foul, respirators or breathing apparatus, as appropriate, should be worn. Due consideration must be given to preventing the onset of 'Weil's Disease', a 'flu-like disease which if untreated can have a serious or fatal outcome. It is transmitted in rat's urine and enters the body through breaks in the skin or, more rarely, by ingestion of contaminated food.

6.3.5 Vibration-induced white finger

The vibrations from portable pneumatic drills and hammers can produce a condition known as white finger or Raynaud's phenomenon in which the tips of the fingers go white and feel numb as if the hand was cold. Anyone showing these symptoms should be taken off work involving the use of these drills or hammers and found alternative employment.

6.3.6 Ionising radiations

There are two main uses for radioactive substances that give off ionising radiations on construction sites. Firstly, tracing water flows and sewers where a low powered radioactive substance is added to the flow and its route followed using special instruments. Only authorised specialists should be allowed to handle the radioactive substance before it is added to the water. Once it is added, it mixes rapidly with the water and becomes so diluted as not to present a hazard.

The second application is in the non-destructive testing of welds where a very powerful gamma (γ) source is used. Because of its penetrating powers and the effects its rays have on human organs, very strict controls must be exercised in its use. The relevant precautions are detailed in Regulations[14] whose requirements must be complied with.

6.3.7 Lasers

Lasers are beams of intense light, they are radiations but do not ionise surrounding matter. Hazards stem from the intensity of the light which can burn the skin, and, if looked into, can cause permanent damage to eyesight. Ideally, lasers of classes 1 or 2 should be used as these present little hazard potential. Class 3A lasers give rise to eye hazards and should only be used in special cases under the supervision of a laser safety adviser. Class 3B and above generally should not be used on construction work, but if the necessity arises only adequately trained persons should operate them. When eye protection is assessed as being necessary, the type supplied must be certified as providing the required attenuation for the laser being used[15].

6.3.8 Compressed air work

The health hazards of work in compressed air and diving are decompression sickness ('the bends') and aseptic bone necrosis ('bone rot'). Both these illnesses can have long-term effects varying from slight impairment of mobility to severe disablement. The protective measures, including decompression procedures, are laid down in the Work in Compressed Air Special Regulations[16] and these should be observed in conjunction with the medical code of practice[17] for work in compressed air. Where diving work is involved, the Diving at Work Regulations 1997[1] apply.

6.4 Access

6.4.1 General access equipment

Although there is a trend in the construction industry towards specialised plant to meet a particular need, the most common material at present employed to provide access scaffolding is scaffold tube and couplers. Large-scale or difficult projects are best carried out by experts but there is a very large amount of scaffold erection of the smaller type in short-term use which can be quickly and safely erected by craftsmen who are to work on them, provided they have been trained in the basic techniques and requirements of the British Standard Code of Practice[18].

As with all structures, a sound foundation is essential. Scaffolds must not be erected on an unprepared foundation. If soil is the base it should be well rammed and levelled and timber soleplates at least 225 mm (9 in) wide and 40 mm (1½ in) thick laid on it so that there is no air space between timber and ground.

The standards should be pitched on baseplates 150 mm × 150 mm (6 in × 6 in) and any joints in the standards should occur just above the ledger. These joints should be staggered in adjacent standards so that they do not occur in the same lift. Ledgers should be horizontal, placed inside the standards and clamped to them with right-angle couplers. Joints should be staggered on adjacent ledgers so that they do not occur in the same bay.

Decking will generally be 225 mm × 40 mm (9 in × 1½in) boards and each board should have at least three supports but this is dependent upon the grade of timber used for the boards. The British Standard[19] recommends that they do not exceed 1.2 m (4 ft). Boards are normally butt jointed but may be lapped if bevel pieces are fitted or other measures taken to prevent tripping. A 40 mm (1½ in) board should extend beyond its end support by between 50 mm and 150 mm (2 in and 6 in).

Guardrails must be fitted at the edges of all working platforms at a height of at least 910 mm with an intermediate guardrail so there is no opening greater than 470 mm between any guardrail or toe board. An alternative to an intermediate guardrail is the use, between the top guardrail and the decking, of in-fill material which should be of sufficient strength to prevent a person from falling through the gap.

Ladders must stand on a firm level base and must be secured at the top and bottom so that they cannot move. All ladders must extend at least 1.07 m (3 ft 6 in) beyond the landing level. To preserve them they may be treated with a clear preservative or be varnished but must not be painted. All rungs must be sound and properly secured to the stile. No ladder, or run of ladders, shall rise a vertical distance exceeding 9 m unless suitable and sufficient landings or rest areas are provided.

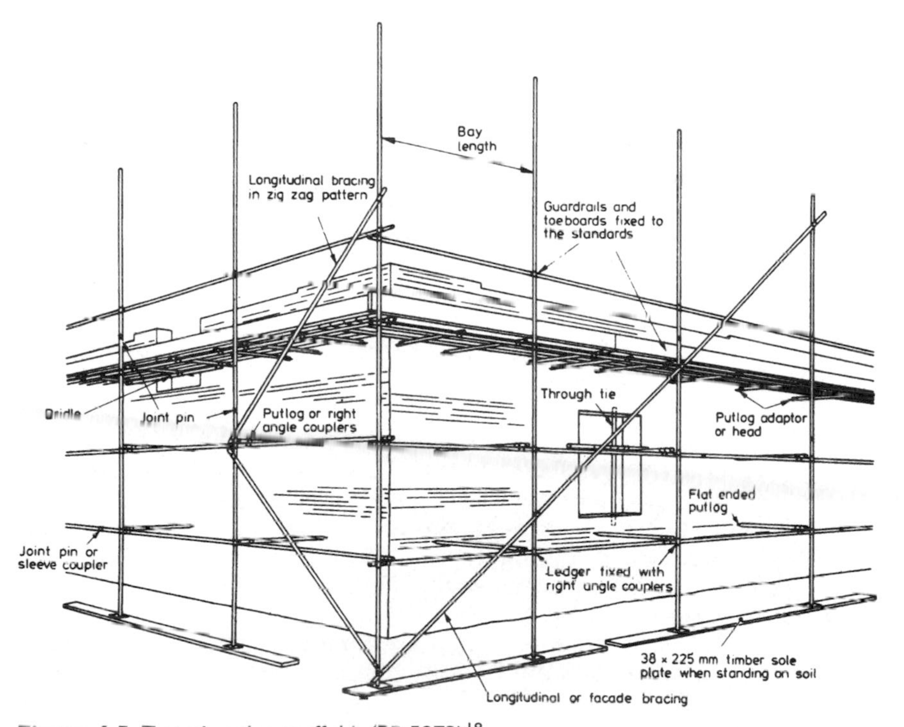

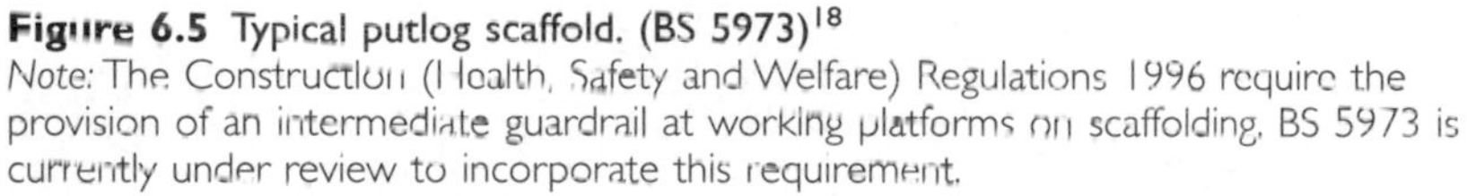
Figure 6.5 Typical putlog scaffold. (BS 5973)[18]
Note: The Construction (Health, Safety and Welfare) Regulations 1996 require the provision of an intermediate guardrail at working platforms on scaffolding. BS 5973 is currently under review to incorporate this requirement.

Unless properly designed to stand on their own, all scaffolds must be sufficiently and effectively anchored to the building or structure by ties which are essential to ensure stability of the scaffold. Before using a scaffold, the employer has a duty to arrange for it to be inspected by a competent person, then ensure that it is inspected every seven days and a record maintained of the inspection. All scaffold material must be kept in good condition and free from patent defect. Damaged equipment should be stored separately and identified as 'damaged' or destroyed. Metal scaffold tubes and fittings and timber scaffold boards should comply with the appropriate British Standard[19,20]. *Figure 6.5* shows a simple scaffolding structure where only an outside row of standards are used to support the platforms, with putlogs fixed into brickwork joints.

6.4.2 Mobile towers

A mobile access tower[21] (*Figure 6.6*) is a tower formed with scaffold tube and mounted on wheels. It has a single working platform and is provided with handrails and toeboards. It can be constructed of prefabricated tubular frames and is designed to support a distributed load of 30 lb/ft^2. The height of the working platform must not exceed three times the smaller base dimension and no tower shall have a base dimension less than 4 ft. Rigidity of the tower is obtained by the use of diagonal bracing on all four elevations and on plan. Castors used with the tower should be fixed at the extreme corners of the tower in such a manner that they cannot fall out when the tower is moved and shall be fitted with an effective wheel brake. When moving mobile towers great care is essential. All persons, equipment and materials must be removed from the platform and the tower moved by pushing or pulling at the base level. Under no circumstances may mobile towers be moved by persons on the platform propelling the tower along.

6.5 The Construction (Lifting Operations) Regulations 1961

These Regulations apply to all equipment used for lifting on construction sites and include fixed, mobile and travelling cranes, hoists used for both goods and passengers and also the ropes, chains, slings etc. that support the load being lifted. These Regulations are due to be replaced by the Lifting Operations and Lifting Equipment Regulations 1998 (LOLER) which offer the option of schemes for inspections which are either risk based or prescribed.

There are certain requirements that are common to all this equipment in that they must be of sound construction and free from patent defect (regs. 10, 21, 34, 47). When erected the equipment must be properly supported and secured (regs. 11, 18) and that ground conditions are such as to ensure stability (regs. 19, 20). Erection must be under competent control (reg. 25) and in the case of derricking cranes follow certain procedures (reg. 33).

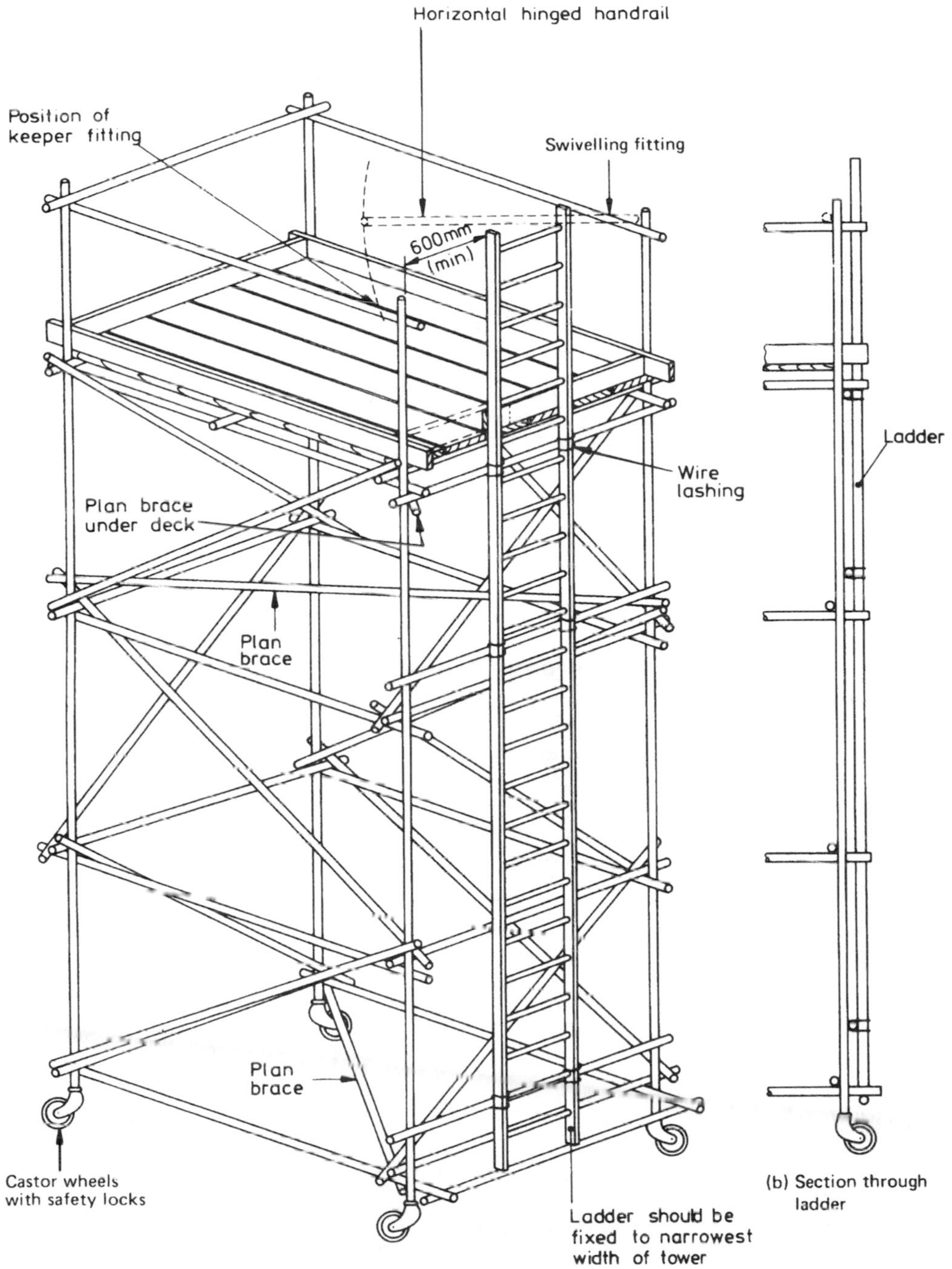

Figure 6.6 Mobile access tower. (BS 5973)[21]

Note: The Construction (Health, Safety and Welfare) Regulations 1996 require the provision of an intermediate guardrail at working platforms on scaffolding. BS 5973 is currently under review to incorporate this requirement.

All lifting equipment must be examined regularly (regs. 10, 28, 34, 46) with safe means of access provided for those carrying out the inspection (reg. 17). The safe working load must be clearly indicated (regs. 29, 45) and jib cranes must have an automatic safe load indicator (reg. 30) which must be tested. The specified safe working load must not be exceeded (reg. 31).

The positioning of travelling or slewing cranes should be such that a clear passageway, 0.6 m (2 ft), is ensured at all times (reg. 12). Where drivers or banksmen require platforms that are more than 2 m (6 ft 6 in) above an adjacent level, suitable guardrails must be provided (reg. 13) and a cab with safe access should be provided for drivers exposed to the weather (reg. 14). Any communication between banksman and driver must be clear (reg. 27). *Figure 3.19* (p. 4.101) shows the visual signals in common use[22].

Lifting tackle such as chains, rings, hooks, shackles etc. must not be modified by welding unless by a competent person and followed by a test (reg. 35). Hooks should have a safety clip (reg. 36) and slings must not be used in such a way that is likely to damage them (regs 37, 38, 39). Chains, ropes and lifting gear must be thoroughly examined by a competent person every six months (reg. 40) and chains and hooks of wrought iron are to be heat-treated every 14 months (reg. 41).

A hoist is defined (reg. 4) as '. . . a lifting machine, whether worked by mechanical power or not, with a carriage, platform or cage the movement of which is restricted by a guide or guides, but does not include a lifting appliance used for the movement of trucks or wagons on a line or rails' and special requirements attach to the protection of hoistways, platforms and cages (reg. 42) and to their operation (reg. 43). Special precautions apply where people are carried by lifting equipment (Regs. 47, 48) and to the securing of loads (reg. 48).

Records must be kept of the examinations of lifting equipment (reg. 50).

To facilitate compliance with these requirements, checklists can be used for the different items of lifting gear and tackle and the following are typical lists that apply to the various statutory requirements under this section of the Construction Regulations.

6.5.1 Checklists

6.5.1.1 Mobile cranes

Prior to work commencing ensure a competent 'lifting co-ordinator' has been appointed:

When was the crane selected, and what information was available/used at the time?

Has the selected crane been supplied?

Check that the ground is capable of taking the loads (outriggers/crane/load/wind). If in doubt get ADVICE from specialist departments/firms.

Ensure that the approach and working area are as level as possible.

Ensure that the area is kept free of obstructions – minimum 600 mm (2 ft) clearance.

Ensure that the weights of the loads are known, and that the correct lifting gear is ordered/available.
Ensure that there is a competent, trained banksman available.
Check that there are no restrictions on access, i.e. check size(s) of vehicles etc.
Ensure that the work areas are adequately lighted.
Check that the Plant Department/Hirer has provided information re the cranes etc.
Whilst work is in progress:
Check that there is an up-to-date test certificate.
Check that the daily/weekly inspections are being carried out and entered in Form 91 (Part 1).
Ensure that the crane is operating from planned/approved positions only.
Ensure that the banksman is working in the correct manner.
Ensure that the correct lifting gear is being used.
Ensure that outriggers are being used, and are adequately supported.
Check that the safe load/radius indicator is in working order.
Check that the tyres/tracks are at the correct pressure and in good, clean condition.
Check that the crane is kept at a safe, predetermined distance from open excavations etc.
Check that, when travelling, the load is carried as near to the ground as possible and that hand lines are being used.
Check that when travelling on sloping ground the driver changes the radius to accommodate the moving of the load.
Check that loads are not being slewed over persons and that persons are not standing or walking under the load.

6.5.1.2 Automatic safe load indicator

All cranes with a lifting capacity in excess of 1 ton must be fitted with an approved type of automatic safe load indicator (Construction (Lifting Operations) Regulations 1961)

It is the responsibility of the operator to:

(a) determine the type of indicator fitted;
(b) determine how the adjustments are made;
(c) ensure that it is correctly adjusted for the various lifting duties;
(d) ensure that the electrical circuit is tested for serviceability;
(e) take immediate action when an overload is indicated.

The signals given by the indicator take the form of coloured lights, a dial indicator or both and a bell.

Green/white – Indicator adjusted for 'free' duties
Blue – Indicator adjusted for 'blocked' duties
Amber – Maximum safe load being approached
Red – Overload condition reached.

The red light will be supported by a bell to give an audible warning of overload.

6.5.1.3 Goods hoists, static and mobile – safe working checklist

(a) Erect the hoist in a suitable position.
(b) Make the hoistway as compact as possible.
(c) Hoistway to be efficiently protected by a substantial enclosure at least 2 m (6 ft 6 in) high.
(d) Hoist gates – guards to be at least 2 m (6 ft 6 in) high.
(e) Engine or motor must also be enclosed to a height of 2 m (6 ft 6 in) where practicable.
(f) Make sure that no one can come into contact with any moving part of the hoist.
(g) Enclosures at the top may be less than 2 m (6 ft 6 in) but in no case less than 0.9 m (3 ft) providing that no one can fall down the hoistway and that there is no possibility of anyone coming into contact with any moving part.
(h) All intermediate gates will be 2 m (6 ft 6 in) unless this is impractical, i.e. confined space etc.
(i) The construction of the hoist shall be that it can only be operated from the upmost working position at any one time.
(j) It shall not be operated from inside the cage (unless designed for the purpose).
(k) The person operating the hoist must have a clear and unrestricted view of the platform throughout.
(l) The safe working load shall be plainly marked on every hoist platform and this load must not be exceeded.
(m) No person shall ride on the hoist (unless so designed), and a notice to this effect must be exhibited on the hoist so that it can be seen at all levels.
(n) Every hoist must be fitted with an efficient automatic device to ensure that the platform does not overrun the highest point for which it is intended to travel.
(o) Every hoist must be fitted with an efficient device which will support the platform and load in the event of the failure of the ropes or lifting gear.
(p) All movable equipment or plant must be scotched, to prevent its displacement while in motion.
(q) All materials will be so placed as to prevent displacement.
(r) Gates should be kept closed on all landing stages. Every person using hoists must close landing place gates immediately after use. (This is a statutory duty imposed upon the person actually using the hoist but the employer also has the duty of seeing that the regulation is obeyed, and the employer's representative on site is the General Foreman or Site Agent.)
(s) Landing stages should be kept free from materials and plant.
(t) No person under the age of 18 must be allowed to operate or give signals to operator.
(u) Only a competent person should operate the hoist.

(v) Signals to be of distinct character; easily seen or heard by person to whom they are given.
(w) Every hoist must be inspected once a week and a written entry should be made weekly in Form 91 (Part I Section C).
(x) Every hoist should be examined every six months by a competent person and certificated.

6.5.1.4 Chains, rope slings and lifting gear – safe working checklist

Prior to commencing work:

(a) Examine the slings provided and check that the 'thorough examination' has been carried out and recorded.
(b) Determine and clearly mark the Safe Working Loads for all slings.
(c) Ensure that the correct and up-to-date copies of the Sling Chart and Safe Working Load Tables are available, when using multi-leg slings.
(d) Ensure that a copy of the correct crane signals is available.
(e) Ensure that a suitable rack is available for storing slings, etc. not in use. N.B. Wire ropes should be stored in a dry atmosphere.
(f) Ensure that the weights of loads to be lifted are known in advance, and that load weights are clearly marked.
(g) Find out the type of eye bolt fitted to the load, in advance, to ensure that the correct equipment, shackles/hooks/lifting beams, is available on site.

Whilst work is in progress:

(h) Ensure that the 'right' techniques are being used.
(i) Ensure that the copies of the Sling Chart and the Safe Working Loads Tables are being used, where necessary.
(j) Ensure that the correct crane signals are being used, and that signals are given only by 'approved' banksmen.
(k) Ensure that regular inspections of the equipment are being carried out.
(l) Ensure that unfit slings are destroyed, or at least removed from site.
(m) Stop persons 'hooking back' onto the legs of slings.
(n) Ensure that slingers understand that 'doubling up' the sling does NOT 'double up' the Safe Working Load: avoid this practice if possible.
(o) Limit the use of endless wire rope slings.
(p) Ensure that wire rope slings are protected from sharp corners of the loads, by suitable packings.
(q) Prevent/stop slings/ropes from being dragged along the ground.
(r) Ensure that the hooks used for lifting are NOT also carrying unused slings.
(s) Ensure that the crane hook is positioned above the load's centre of gravity.

(t) Ensure that the load is free before lifting and that all legs have a direct load.
(u) Ensure that the load is landed onto battens to prevent damage to slings.
(v) Ensure that a sling is NOT passed through more than one eye bolt.
(w) Ensure that 'snatch' loading does NOT take place.
(x) Ensure that NO ONE rides on a load that is being slung.

Additional guidance on the standards to be achieved is given in British Standards[22–24] and in HSE Guidance Notes[25].

6.6 Welfare facilities

Under the Construction (Health, Safety and Welfare) Regulations 1996 every contractor or employer has a duty to provide, or ensure there is provided, certain health and welfare facilities for his own employees who must have proper access to them. Because a number of contractors may be working on the site some of the facilities may be shared (reg. 4) or alternatively arrangements may be made by the contractor to use the facilities offered by adjacent factories. Such arrangements should be agreed in writing. When such agreements are terminated, to prevent confusion it is advisable to give notice in writing.

If facilities are shared with another employer or contractor on the site, then the one who provides the facilities or equipment must:

1 in deciding what facilities to provide, assume that he employs the total number of men who are to use the facilities: e.g. say own employees = 50, other employees = 40 – therefore, for the purposes of providing facilities, assume that he employs 90;
2 Keep a record showing the facilities to be shared and the names of the firms sharing them.

Contractors or employers on the same site can jointly appoint the same man to take charge of first aid and ambulance arrangements.

6.6.1 Facilities to be provided on site

Clearly marked 'FIRST AID' boxes must be provided and put in the charge of a responsible person whose name must be displayed near the box. After assessing the level of risk, the availability of emergency services and other matters detailed in the Regulations[26] and the Approved Code of Practice[27], it may be necessary for the responsible person to be a trained and certificated first aider.

Where there is a large workforce on a site a suitably staffed and equipped first aid room should be provided. However, where a large workforce is divided into several dispersed working groups or the location of the site makes access to places of treatment outside it difficult, the needs of such a site may be better met by the provision of first aid equipment and trained first aiders at different parts of the site.

Regardless of the number of employees there must be at least one first aid box on site, and provision should be made for every employee to have reasonably rapid access to first aid.

Construction workers are frequently exposed to the weather and facilities must be provided to store the clothes they do not wear while working, to warm themselves, and to dry their clothing when not in use. In addition, a supply of drinking water must be available and suitable arrangements for warming and eating food.

Suitable washing facilities must be provided and toilets, accessible from all workplaces on the site, must be under cover, partitioned from each other, have a door with fastening, be ventilated and provided with lighting. They must not open directly into workrooms or messrooms and must be kept clean. Separate conveniences must be provided for men and women unless each convenience is in a separate room, the door of which can be secured from the inside.

6.7 Other relevant legislation

6.7.1 Personal protective equipment

The requirements to be met in the application and use of this equipment are laid down in the Personal Protective Equipment at Work Regulations 1992[1] (PPER).

The general assessment required by MHSWR should identify the hazards, the extent of the risks faced and enable the necessary preventive and precautionary measures to be decided. If personal protective equipment is considered appropriate, PPER sets out the steps to be taken in the process of selecting suitable and effective equipment which the employer has to provide to his employees.

These Regulations revoke early regulations made under FA 1961, such as the Protection of Eyes Regulations 1974, and have marginally modified subsequent regulations that include provisions for personal protective equipment, such as the Noise at Work Regulations 1989, the Construction (Head Protection) Regulations 1989 etc., which continue to apply.

6.7.2 The Construction (Head Protection) Regulations 1989[1]

Under these Regulations employers are required to provide, maintain and replace, as necessary, suitable head protection for their employees and others working in the areas over which they have control. They must ensure that the head protection is worn unless there is no risk of a head injury occurring other than by falling over. To be 'suitable' head protection must be:

- designed, so far as is reasonably practicable, to provide protection against forceseeable risks of injury to the head to which the wearer may be exposed,
- adjustable so that it can be made to fit the wearer, and
- suitable for the circumstances in which it is to be used.

Persons who have control over a site, such as management contractors, may make written rules concerning the wearing of head protection by anyone, employees and others (with the exception of Sikhs wearing turbans), working on that site, and should make arrangements for enforcing those rules. Employees are required to take care of equipment and to report cases of damage, defect or loss.

6.7.3 Fire Certificates (Special Premises) Regulations 1976

(a) These Regulations were introduced as a result of the operation of the Health and Safety at Work Act with effect from 1 January 1977. Most temporary site buildings used as offices or workshops at building operations and works of engineering construction were previously subject to certain fire precautions provisions, under the Offices, Shops and Railway Premises Act, or occasionally under the Factories Act. Health and Safety Executive Inspectors already inspect building sites for general inspection purposes, and by including these premises into the new Regulations the Health and Safety Executive now becomes responsible for fire precautions at them. It is not intended to issue a fire certificate for every temporary site building, however, and the Regulations are designed to exclude small site buildings conditionally from the certification procedure.

Where a site building, used as an office or workshop, does not require a fire certificate, the Construction (Health, Safety and Welfare) Regulations require certain fire detection and firefighting measures to be implemented similar to those in buildings where construction work is carried out. Steps must be taken to:

1 provide suitable and sufficient firefighting equipment
2 provide suitable and sufficient fire detectors and alarms
3 maintain, inspect, examine and test any equipment provided
4 ensure non-automatic equipment is readily accessible
5 provide training for the workforce.

Some large site buildings, or those with special risks, will require a certificate.

(b) A fire certificate will be required for any building or part of a building which is 'constructed for temporary occupation for the purposes of building operations or works of engineering construction, or, which is in existence at the commencement there of any further such operations'. But a fire certificate will not be required for these buildings if all of the following conditions are complied with:

(i) Not more than 20 persons are employed at any one time in the building or part of the building.
(ii) Not more than 10 persons are employed at any one time elsewhere than on the ground floor.
(iii) No explosive or highly flammable material is stored or used in or under the building.

(iv) The building is provided with reasonable means of escape in case of fire for the persons employed there.
(v) Appropriate means of fighting fire are provided and maintained and so placed as to be readily available for use in the building.
(vi) While anyone is inside, no exit doors may be locked or fastened so that they cannot be easily opened from inside.
(vii) If more than 10 people are employed in the building, any doors opening on to any staircase or corridor from any room in the building must be constructed to open outwards unless they are sliding doors.
(viii) Every exit opening must be marked by a suitable Notice.
(ix) The contents of every occupied room in the building must be arranged so that there is a free passageway for everyone employed in the room to a means of escape in case of fire.

(c) To obtain a fire certificate application must be made to the HSE (Construction) office for the area in which the site is located, using form F2003 to give the required particulars, which are listed below. The HSE will then carry out an inspection of the site buildings. In practice, all final exit doors and any door opening on to a staircase or corridor from any room in the building should open outwards. Fire extinguishers to be hung on wall brackets adjacent to the final exit, with the top of the extinguisher 1.07 m (3 ft 6 in) from the floor. A fire blanket should be hung on wall bracket adjacent to the cooker in the canteen with the top of the blanket 1.52 m (5 ft 0 in) from the floor. LPG cylinders should be sited outside huts and have an isolating valve at the cylinder and an ON/OFF control valve as near as practicable to the heater.

Particulars required on form F2003 when applying for a fire certificate include:

(a) Address of premises.
(b) Description of premises selected from those listed in Schedule 1 to the Regulations.
(c) Nature of the processes carried on, or to be carried on, on the premises.
(d) Nature and approximate quantities of any explosive or highly flammable substance kept, or to be kept, on the premises.
(e) Maximum number of persons likely to be on the premises at any one time.
(f) Maximum number of persons likely to be in any building of which the premises form part at any one time.
(g) Name and address of any other person who has control of the premises.
(h) Name and address of the occupier of the premises.
(j) If the premises consist of part of the building, the name and postal address of the person or persons having control of the building or other part of it.

6.7.4 Food Safety (General Food Hygiene) Regulations 1995

The Food Safety Act 1990 is now one of the main pieces of legislation concerning food hygiene. It is an enabling Act allowing specific subordinate legislation (Regulations) to be made as necessary. An example is the Food Premises (Registration) (Amendment) Regulations 1993 which require all places where food is served, including site canteens, to be registered with the local authority. The 1995 Regulations establish a set of rules within which businesses handling and preparing food must operate. All food premises, particularly food rooms, must be arranged so that surfaces can be cleaned properly, waste disposed of hygienically and a high standard of personal hygiene maintained by the food handlers. The proprietor must have a system which identifies the points in the food handling process where contamination could occur and have in place control measures to avoid the risk. He can achieve this by following the principles of the Hazard Analysis and Critical Control Points (HACOP) outlined in Schedule 2 to the Regulations.

If a person handling food suffers a specified infection of the digestive system, the Medical Officer of Health must be notified immediately. An adequate supply of hot and cold water and hand wash basins must be supplied for the use of food handlers, and separate sinks with hot and cold water for preparing vegetables and for washing equipment. No toilet may connect direct with a room used for preparing or eating food. Food rooms must be adequately lit and ventilated and walls, floors, doors, windows, ceilings, woodwork etc., must be kept clean and in good repair. Waste must not be allowed to accumulate.

6.7.5 Petroleum Consolidation Act 1928 Petroleum Spirit (Motor Vehicles etc.) Regulations 1929 Petroleum Mixtures Order 1929

Broadly, the Act and Regulations deal with petroleum spirit and mixtures whether liquid, viscous or solid. Petroleum means substances giving off a flammable vapour at a temperature of less than 23°C (73°F). It will be appreciated that this is less than body heat so there is a need for strict control on their uses. By prior arrangement with the local authority, up to 60 gallons may be kept on site without licence but this is subject to a review annually. Where quantities in excess of 60 gallons are to be kept, a petroleum licence must be obtained from the Petroleum Officer. These requirements are subject to discussion on their repeal.

6.7.6 Mines and Quarries Act 1954

The requirements of this Act apply to sites, or parts of sites, that fall within the description of a quarry, as defined in s. 180 of the Act, the most common example being a borrow pit used in motorway construction, although it can also apply where old quarries are being in-filled with spoil. Enforcement of this Act is split between the Factory Inspectorate who cover building on the surface, and the Mines and Quarries Inspectorate who deal with work within the quarry itself.

Notification of the commencement and termination of work in a quarry must be made on Mines and Quarries form 213. The owner of the quarry is required to appoint a manager and a deputy manager.

6.7.7 Asbestos

Asbestos is a strong, durable, non-combustible fibre, and these physical properties make it ideal as a reinforcing agent in cement, vinyl and other building materials, e.g. vinyl floor tiles, bath panels, cold water cisterns, roof felts, corrugated or flat roof sheets, cladding sheets, soffit strips, gutters and distribution pipes. It is also a good insulant and has been used for protecting structures from the effects of fire.

Asbestos has been used extensively in the past and may be found in many forms in existing buildings including: industrial wall and roof linings; internal partitions; duct and pipe covers; suspended ceilings; fire doors and soffits to porch and canopy linings. It also occurs in plant rooms and boiler houses and as asbestos coatings and insulating lagging on structures and pipework.

Work with asbestos products is governed in the UK by specific legislation:

The Asbestos (Licensing) Regulations 1983
The Asbestos (Prohibitions) Regulations 1992
The Asbestos Products (Safety) Regulations 1985
The Control of Asbestos at Work (Amendment) Regulations 1992
The Control of Asbestos in the Air Regulations 1990

The main purpose of the 1983 Regulations is to control the manner in which companies and the self-employed carry out work which involves the disturbance of asbestos insulations or coatings. All persons carrying out such work must be in possession of an HSE licence and receive regular medical examinations. Wherever asbestos products are encountered the Regulations should be complied with, using advice given in HSE guidance notes[28]. Useful information is also available from the Asbestos Information Centre Ltd.

Control limits for asbestos dust in the working environment are set out in an Approved Code of Practice[29]. These control limits are not intended to represent safe levels of airborne dust but the upper limits of permitted exposure. There is still a statutory duty to reduce exposure to the lowest level that is reasonably practicable.

Where asbestos containing materials have to be stripped, its disposal is governed by special regulations[30].

References

1. Statutory Instruments relevant to construction work:
The Construction (Lifting Operations) Regulations 1961 (SI 1961 No. 1581)
The Electricity at Work Regulations 1989
The Reporting of Injuries, Diseases and Dangerous Occurrences Regulations 1995

The Highly Flammable Liquids and Liquefied Petroleum Gases Regulations 1972 (SI 1972 No. 917)
The Diving at Work Regulations 1997
Work in Compressed Air Regulations 1996
Noise at Work Regulations 1989 (SI 1989 No. 1790)
Construction (Head Protection) Regulations 1989 (SI 1989 No. 2209)
Lifting Plant and Equipment (Records of Test and Examination etc.) Regulations 1992 (SI 1992 No. 195)
The Management of Health and Safety at Work Regulations 1992 (SI 1992 No. 2051)
The Provision and Use of Work Equipment Regulations 1992 (SI 1992 No. 2932)
The Manual Handling Operations Regulations 1992 (SI 1992 No. 2793)
The Personal Protective Equipment at Work Regulations 1992 (SI 1992 No. 2966)
The Workplace (Health, Safety and Welfare) Regulations 1992
Confined Spaces Regulations 1997 (SI 1997 No. 1713)
All published by HMSO, London.

2. Health and Safety Executive, Manufacturing Services and Industries – Annual Report (published each year), HSE Books, Sudbury
3. Health and Safety Executive, *Blackspot Construction, A study of five years fatal accidents in the building and civil engineering industries*, HSE Books, Sudbury (1988)
4. Health and Safety Commission, *The Construction (Design and Management) Regulations 1994*, HMSO, London (1994)
5. Regina *v.* Swan Hunter Shipbuilders Ltd and Telemeter Installations Ltd [1981] IRLR 403
6. British Standards Institution, BS 6031, *Code of Practice for earthworks*, BSI, London (1981)
7. Health and Safety Executive:
   Health and Safety Guidance Booklet No. HS(G)47, *Avoiding danger from underground services*, (1989)
   Health and Safety Guidance Note No. GS6, *Avoiding of danger from overhead electrical lines*, (1991)
   HSE Books, Sudbury
8. Health and Safety Executive, *Construction (Lifting Operations) Regulations 1961 – Certificate of Exemption: CON(LO)/1981/2 (General), Excavator, loaders and combined excavator/loaders used as cranes*, HMSO, London (1981)
9. Health and Safety Executive, *Health and Safety Guidance Note No. GS7, Accidents to children on construction sites*, HSE Books, Sudbury (1989)
10. BS 5304 Code of Practice for safety of machinery, British Standards Institution, London (1988)
11. Health and Safety Executive, Guidance Notes:
    No. MS 15 Welding
    No. PM 17 Pneumatic Nailing and Stapling Tools
    No. PM 14 Safety in the Use of Cartridge Operated Tools
    No. PM 5 Automatically Controlled Steam and Hot Water Boilers
    No. PM 1 Guarding of Portable Pipe Threading Machines
12. *The Electricity at Work Regulations 1989*, HMSO, London (1989)
13. *The Control of Substances Hazardous to Health Regulations 1994*, HMSO, London (1994)
14. *The Ionising Radiations Regulations 1985*, HMSO, London (1985)
15. British Standards Institution, BS 7192, *Radiation, Safety of laser products*, BSI, London
16. *The Work in Compressed Air Special Regulations 1958*, HMSO, London (1973)
17. *Medical Code of Practice for Work in Compressed Air*, Report No. 44, 3rd edn, CIRIA, London (1988)
18. BS 5973, *Code of Practice for access and working scaffolds and special scaffold structures in steel*, British Standards Institution, London
19. BS 2482, *Specification for timber scaffold boards*, British Standards Institution, London
20. BS 1139 – Metal scaffolding
    Part 1: *Specification for tubes for use in scaffolding*
    Part 2: *Specification for couplers and fittings for use in tubular scaffolding*
    Part 4: *Specification for prefabricated steel splitheads and trestles* British Standards Institution, London
21. Health and Safety Executive, Health and Safety Guidance Notes No. GS 42, Tower Scaffolds, HSE Books, Sudbury (1987)

22. BS 5744, *Code of Practice for safe use of cranes (overhead/underhung travelling and goliath cranes, high pedestal and portal jib dockside cranes, manually operated and light cranes, container handling cranes and rail mounted low carriage cranes)*, British Standards Institution, London
23. CP 3010, *Safe uses of cranes (mobile cranes, tower cranes and derrick cranes)*, British Standards Institution, London (1972) (see also ref. 24)
24. BS 7121, parts 1–5, *Code of Practice for Safe use of cranes*, British Standards Institution, London
25. Health and Safety Executive, Guidance Notes:
PM 3 Erection and Dismantling of Tower Cranes
PM 9 Access to Tower Cranes
PM 8 Passenger Carrying Paternosters
PM 16 Eyebolts
PM 27 Construction Hoists
GS39 Training of Cranes Drivers and Slingers,
HSE Books, Sudbury
26. *The Health and Safety (First Aid) Regulations 1981*, HMSO, London (1981)
27. Health and Safety Executive, Legal series booklet No. L74, First aid at work.
Health and Safety (First Aid) Regulations 1981. Approved Code of Practice and Guidance, HSE Books, Sudbury (1997)
28. Health and Safety Executive, Guidance Notes:
EH35 Probable asbestos dust concentrations at construction processes
EH36 Work with asbestos cement
EH37 Work with asbestos insulating board
EH40 Occupational exposure limits,
HSE Books, Sudbury
29. Health and Safety Executive, Legislation booklet L27, The control of asbestos at work. The Control of Asbestos at Work Regulations 1987, Approved Code of Practice, HSE Books, Sudbury (1988)
30. *The Environmental Protection Act 1990*,
The Special Waste Regulations 1996,
HMSO, London

Further reading

Dickie, D.E., *Crane Handbook*, (Ed.: Douglas Short), Butterworth, London (1981)
Dickie, D.E., *Lifting Tackle Manual*, (Ed.: Douglas Short), Butterworth, London (1981)
Construction Safety Manual, Construction Safety, Crawley, Sussex
Dickie, D.E., Ed. Hudson, R., *Mobile Crane Manual*, Butterworth, London (1985)
King, R. and Hudson, R., *Construction Hazard and Safety Handbook*, Butterworth, London (1985)

Chapter 7

Managing chemicals safely

John Adamson

7.1 Introduction

All chemicals can be handled safely! However, the skill is to know the nature and hazardous properties of the materials so that suitable precautionary arrangements can be made for their safe use and handling. It is also necessary to know where the materials are to be used, by whom and for what. Consideration needs to be given to the effects that escapes of the material can have on the community and the ecology.

A chemical incident has the potential to affect many people, the operators, other employees on site, members of the public and, in the wider sense, it can cause widespread damage to the ecology. The results of some incidents such as at Seveso and Bhopal have been of epic proportions and were truly catastrophies. Such events carry an massive price tag in terms of human suffering, loss of life, business interruption and damage to the environment. All incidents, regardless of their scale, damage the reputation of the organisation and of the industry as a whole.

7.2 Chemical data

Two keys to the safe use and handling of hazardous materials are, first, to know what substances are and, second, to know the characteristics of those substances. The Control of Substances Hazardous to Health Regulations 1996[1] (COSHH) require that data on each of the hazardous substances on the site should be kept and be available to operators at all times. A supporting Approved Code of Practice[2] recommends that a list of all hazardous materials is also kept and made available. Hazardous substances are defined in COSHH as:

> 'Those substances specified as very toxic, toxic, corrosive or irritant within the meaning of the Chemicals (Hazard Information and Packing for Supply) Regulations 1994[3] (CHIP 2) as amended in 1996 (CHIP 96).'

The details are given in an associated Approved Supply List[4]. In essence, the substances with any of the following properties are hazardous to some degree:

(a) explosive
(b) oxidising
(c) flammable
(d) highly flammable
(e) extremely flammable
(f) toxic
(g) very toxic
(h) harmful
(i) corrosive
(j) irritant
(k) carcinogenic (causes cancer)
(l) mutagenic (causes inherited changes)
(m) teratogenic (causes harm to the unborn)
(n) micro-organisms that create a hazard to health
(o) substantial concentrations of dust
(p) radioactive materials
(q) any substance not mentioned above which creates a comparable hazard.

7.3 Source of information

Under s. 6 of HSW, manufacturers, importers, designers and suppliers must ensure that articles and substances supplied for use at work are safe and without risk to health. They also have a duty to provide adequate safety information about the substances they produce, which they usually do in the form of chemical safety data sheets. Some employers use these data sheets as they stand, others prefer to generate their own in a format that matches their in-company documentation. Guidance on the information a chemical safety data sheet should contain is given in an HSE publication no. L62[5]. For many of the commonly used substances comprehensive information can be obtained from reference books[6–8]. Where chemicals used in a process are modified during the process, information should also be available concerning the intermediates produced.

Where no hazard data are available for a new substance and it is not possible to estimate its characteristics from substances of similar molecular structure, that substance must be treated as toxic until proved otherwise. Before new substances are put on the market, they must be notified under the Notification of New Substances Regulations 1993 (NONS) with its supporting guide[9] and some may need to be notified under the Chemical Weapons Act 1996. The purpose of NONS is to ensure that adequate technical information is available on new substances before they are put on the market so that suitable precautionary measures can be developed to protect employees, the public and the environment from possible ill effects. NONS support an EU-wide system of notification of new substances.

A new substance is defined as one which is not listed in the European Inventory of Existing Commercial Chemical Substances (EINECS)[10] and may be a substance in its own right or part of a preparation. After notification, new substances are entered in the European List of Notified Chemical Substances (ELINCS)[11]. Both lists give CAS numbers, IOPAC chemical names and some trade names and references.

The Department of Trade and Industry (DTI) is the lead body in the UK for the Chemical Weapons Convention (CWC), which, while primarily an arms control treaty, has implications for both industry and academia since many of the chemicals concerned are dual use goods, i.e. they have legitimate peaceful uses as well as possible military applications. Companies should declare those chemicals which they either produce or use and which are classified under this Convention. A list of those substances can be obtained from the DTI

7.4 Risk assessments

As part of their efforts to ensure a safe working environment, employers are required by the Management of Health and Safety at Work Regulations 1992 (MHSW) to make suitable and sufficient assessments of the health and safety risks arising from their operations as they may affect their employees and others. The assessment should extend to include such aspects as the way in which the work is organised, safety of the product, environmental effects on the local community and ecology.

The object of a risk assessment is to identify all hazards or potential hazards arising from the work so that precautions can be taken to prevent injury or damage to health of employees or anyone else who might be affected.

7.4.1 Definitions

The words 'hazard' and 'risk' arise frequently and are often misunderstood. However, they do have precise meanings and should not be confused:

HAZARD – is the inherent property of a substance to cause harm.
RISK – is a combination of the probability of the hazard causing harm and the severity of the resulting harm or damage.
RISK ASSESSMENT – is a comprehensive quantitative or qualitative evaluation of the probability and degree of possible injury or damage to health from identified hazards with a view to implementing preventive measures.

A risk assessment takes account of all the significant factors that can affect the chance and extent of harm. It should conclude on action needed to manage the risk for the benefit of employees and the company.

7.4.2 Carrying out a risk assessment

Risk assessments need to be carried out by experienced and competent people who are knowledgeable about the hazards of the substances, the equipment and the operations being reviewed. It is a subjective process which cannot normally be validated mathematically, and is essentially practical. Those involved in the assessment need to think logically and laterally, look beyond the obvious and have good interactive skills to tease out crucial information.

Risk assessments for a small unit or department can be carried out by one person but for larger and more complex areas it may be appropriate to involve a small team of people, e.g. safety adviser, area engineer, production team leader, and an operator.

General risk assessment process

To ensure that all facets that could be affected by the hazards are considered a risk assessment should follow a series of logical steps based on a thought-out strategy:

1 Define the task or process to be assessed and identify the boundaries.
2 Identify the hazards and eliminate or reduce them as far as possible.
3 Evaluate the risks from the residual hazards by:
  (a) assessing the extent of the hazard
  (b) estimating the probability of harm occurring
  (c) assessing likely extent of the harm or injury.
4 Decide on precautions or control measures to be taken.
5 Train local operators.
6 Implement precautionary measures.
7 Monitor the effectiveness of the measures and adjust as necessary.

By giving empirical values to items 3(a), (b) and (c) and multiplying them together an indicative risk rating can be obtained which can be used to determine priorities for action.

When carrying out a risk assessment all types of hazard should be considered – injury, fire, explosion, pollution, damage to neighbours, damage to equipment etc. A typical risk assessment form is shown in *Figure 7.1(a)*.

7.4.3 COSHH assessment

An assessment under COSHH is concerned only with hazardous substances and follows the same procedure outlined above. Its aim is to assess the risks to the health of employees from the various substances used or handled. The assessor should be knowledgeable in the process and in the likely health effects of the substances. The assessment should include not only the process operators but those working in the vicinity who could be affected.

Site:		Area:
Operations covered by this Assessment:		
Maximum No. of people exposed: Frequency & durations of exposure:		
<u>Hazards</u>		
<u>Actions already taken to reduce the risk</u>		
Assessment of residual risk:		
Further actions required:		
Signed: Position:	Date:	Review Date

Figure 7.1(a) Risk assessment form. (Courtesy British Sugar plc)

The technique requires:

- Knowledge of the hazards associated with the substance in order to determine the potential of the hazard to harm health.
- Assessment of the level of exposure to, or contact with, the hazardous substance, i.e. the RISK.
- An estimate of the frequency and duration of exposure.
- A comparison of the level of exposure against the current occupational exposure standard (OES) or maximum exposure limit (MEL) for inhalation and absorption risks.

BRITISH SUGAR PLC

COSHH ASSESSMENT AREA: SITE:

SUBSTANCES/MAIN COMPONENTS		STOCK CODE:
TASK/EXPOSURE POINTS	HAZARD:	EH 40 EXP LIMIT:
FREQUENCY/DURATION OF EXPOSURE		NO. OF PEOPLE EXPOSED:
CONTROL MEASURES	W. I. REFERENCES:	
MONITORING REQUIRED: YES/NO	RISK: LOW MEDIUM HIGH	
ACTION REQUIRED	RESPONSIBILITY	BY

SIGNATURE: DATE FOR NEXT ASSESSMENT:

Figure 7.1(b) COSHH assessment form. (Courtesy British Sugar plc)

- Assessment of the numbers of workpeople who may be exposed.
- Agreement for the control measures to be adopted.

By using a numerical rating a measure of the priority ranking for remedial action can be made. In cases where the level of exposure to a material is not clear cut, an occupational hygiene study should be carried out. *Figure 7.1(b)* is an example of the documentation used to record the findings of a COSHH assessment.

7.4.4 Manual handling risk assessment

In the use of chemicals it may be necessary to handle them manually, often in sacks weighing 25 kg or more. Where this occurs, the risks from the manual handling should be included in the risk assessment. This is also a requirement of the Manual Handling Operations Regulations 1992. This assessment should be carried out by a person trained in handling techniques who should consider:

- whether the task can be done mechanically
- the load – its weight, shape, condition, centre of gravity
- the working environment – condition of the floor, lighting, temperature etc.
- the individual's physique and capacity to carry out the task
- the repetitive nature of the task
- other factors such as clothing and personal protective equipment.

Where the task cannot be done mechanically, the risks identified should be reduced as far as possible, such as by dividing the load or getting assistance, to prevent injuries from lifting, repetitive strains and poor posture.

7.5 The management of risk

Having identified the hazards and assessed the risks the action needed to prevent injury and harm should be agreed with the manager and operators of the area concerned and implemented. While nothing can be absolutely and unequivocally safe and free from risk, the aim must be to achieve, so far as is reasonably practicable (see *Edwards* v. *National Coal Board*[12]), a standard that reduces risks to a minimum whilst maintaining the viability of the process.

Within the strategy outlined in section 7.4.2 above, there are a number of techniques that can be used to ensure the hazards, and consequently the risks, are reduced to a minimum.

7.5.1 Prevention of exposure

With hazardous materials, the first and most basic question that must be asked is 'Do we have to use this substance?' Depending on the answer, there are a hierarchy of facets to consider before deciding on the action necessary to ensure safe working:

(a) *Substitution*.
If a substitute can be found it must meet the following criteria:

- (i) be safer than the original material
- (ii) be suitable for the process to ensure the required quality of finished product – check with chemist and process controller

(iii) be economically viable
(iv) not produce intermediates or daughter products that nullify the benefits of substitution.

(b) *Minimise the quantity*
Ensure that only the minimum quantity of the substance is used in the process. This has the additional benefit of reducing wastage to a minimum. Also the plant should be designed and maintained to minimise the generation of dust, vapours, micro-organisms etc. and, in case of spillage or leakage, be provided with an area for containment.

(c) *Total enclosure*
Total enclosure will ensure that escapes of gases, fumes or liquids are completely contained. Any vents from the enclosure should be through scrubbers or absorbers. The enclosed space may need to be provided with cooling.

(d) *Partial enclosure with local ventilation*
If access is necessary to the plant during parts of the process, local air extraction should be provided and arranged so its exhaust is scrubbed or absorbed then vented to a safe location away from air intake ducts and work areas. For infrequent brief visits, PPE may be an option.

(e) *Local exhaust ventilation* (LEV)
Where regular attention to the process is necessary LEV should be provided. The system must ensure that the hazardous materials do not enter the work area beyond the particular process plant. A laminar flow booth may be a more suitable alternative. Discharge from either system must be through scrubbers or absorbers with final outlet at a safe height and location not to affect neighbours or cause environmental contamination.

(f) *Personal protective equipment* (PPE)
The Personal Protective Equipment at Work Regulations 1992 require employers and the self-employed to provide suitable protective equipment to operators wherever the risks to their health cannot be controlled by other means. PPE should only be used as a last resort and only if engineering control measures are insufficiently effective. PPE should:

- be suitable for the hazardous material involved and the environmental conditions likely to be met
- fit properly, be acceptable to and effective on the wearer allowing for his/her state of health
- not interfere with the work to be carried out by the wearer
- be of an approved type
- not generate further hazards.

A range of PPE, that meets the above criteria, should be tried out by the plant operators before deciding on a particular type. All PPE must be properly looked after and maintained in good working condition. Ideally

PPE should be an individual issue, but if this is not possible it must be cleaned between uses and operators instructed accordingly.

Work instructions should detail the type of PPE to be worn when carrying out specific tasks and operators should receive adequate information and training in its use and limitations. With respiratory protective equipment (RPE) each operator should be 'fit tested' during which they make a number of facial movements in an atmosphere of strong saccharin. If they can taste the saccharin, other more suitable RPE should be tried. Facial hair can also cause failure of the 'fit test'. The testing should be repeated for each type of RPE that has to be worn. *Figure 7.2* shows an operator wearing adequate RPE to protect against chemical dust inhalation.

Figure 7.2 Protective clothing to provide protection against contact and dust. (Courtesy Rhone-Poulenc-Rorer)

The Personal Protective Equipment at Work Regulations do not apply to work areas covered by the more specific requirements of:

The Control of Lead at Work Regulations 1980
The Ionising Radiations Regulations 1985
The Control of Asbestos at Work Regulations 1987
The Control of Substances Hazardous to Health Regulations 1994
The Noise at Work Regulations 1989
The Construction (Head Protection) Regulations 1989

7.5.2 Training and information

An essential feature of any risk management system involving hazardous substances is the training provided to those who handle or use those substances. Such training should include:

(a) Chemical data and information on the substances involved, their hazards and the action to be taken should contamination occur.
(b) The system of work to be followed to ensure that:
- the correct PPE is worn
- the substance is handled properly
- the material is transported safely
- the finished product can be handled safely.

(c) Techniques for dealing with spillages
(d) The operation of special plant or equipment and how to deal with deviations from normal operation or performance.
(e) The purpose and correct use of PPE, its limitation of use, cleaning, care and the reporting of any damage to or faults with any item.
(f) Monitoring techniques and comparison of results with the limits set in EH 40[13]
(g) Checking of the concentrations of substances in finished products and waste.

7.5.3 Monitoring

Monitoring of the atmospheric concentrations to which an employee is exposed when undertaking a task should be carried out where:

(a) The COSHH assessment indicates the level of risk warrants it.
(b) The effectiveness of the corrective environmental control measures in the work area need to be evaluated.
(c) There is a need to demonstrate that the MEL has not been exceeded.
(d) Confirmation is needed that exposure levels are below the appropriate OES.
(e) The effectiveness of the LEV needs to be assessed.

Where an occupational hygiene study has been carried out to check on the effectiveness of the measures to prevent operator contact with hazardous substances, the results should be shared with the operators. Where the exposure risk warrants it, COSHH requires that exposed employees be offered medical surveillance. COSHH also requires that LEV systems are regularly monitored for their effectiveness in controlling the emissions into the work area and that the system still operates within the original design parameters.

7.5.4 Records

COSHH assessment records and occupational hygiene reports should be kept for at least 40 years. These records should include the name and works number of the employees who have been monitored. Health surveillance records should be kept for 5–40 years, dependent upon the substance involved and the degree of exposure.

7.6 Legislative requirements

Following the Flixborough incident[14] which had a major impact on the local community and caused considerable damage to property, the HSE appointed a committee of experts, the Advisory Committee on Major Hazards, to consider the health and safety problems posed by major chemical sites, and to make recommendations. This they did in three reports which identified a need for three basic elements of control:

(i) Identification of the site.
(ii) The location of the site.
(iii) An assessment of the potential hazards on the site.

It had been the intention to implement these three recommendations through a single set of regulations. However, following two major incidents at Seveso and Manfredonia in Italy in 1976 the EU adopted a directive on major hazards. To incorporate this directive into UK law resulted in the preparation of two slightly conflicting sets of regulations, the Notification of Installations Handling Hazardous Substances Regulations 1982 (NIIHHS) and the Control of Industrial Major Accident Hazards Regulations 1984 (CIMAH) with its subsequent amendments. In addition, continuing concern about the effects on workers of exposure to chemicals resulted in, *inter alia*, COSHH and CHIP. Currently, new regulations[15] are in train to incorporate into UK law the contents of an amended EU directive which addresses some current anomalies and expands the type of installation that comes within its scope.

7.6.1 The Notification of Installations Handling Hazardous Substances Regulations 1982 (NIIHHS)

The main thrust of these Regulations is to identify those sites which handle or store more than specified quantities of hazardous substances and it has resulted in the compilation of a central register of all major chemical and other potentially dangerous sites. The Regulations require only that the sites be notified and place no further obligations on employers. Guidance on these Regulations is given in an HSE publication[16].

7.6.2 The Control of Industrial Major Accident Hazards Regulations 1984 (CIMAH)

These are the principal Regulations that govern major hazard sites in the UK and incorporate requirements contained in directive no. 82/501/EEC[17] (often referred to as the Seveso Directive). The Regulations place duties on the owners of hazardous sites to demonstrate safe operation and to notify any major accidents that occur. Where the more dangerous activities, listed in the various schedules, occur the sites have to prepare a 'Safety Case' which describes their activities and their likely impact on the surrounding area. Further, the owners have to prepare an on-site emergency plan and provide information to the local authority who are themselves required to prepare an off-site emergency plan. Finally, information has to be given to the local population who may be affected by the site operations. Guidance on the application of these Regulations is given in two HSE publications[18,19].

These Regulations had a significant impact on the chemical industry in general causing it to be much more overt in its operations. An amendment in 1990 widened the application of the Regulations to include aggregated amounts of substances rather than relating to single substances only. Thus, if a site contains a total of more than 200 tonnes of various substances, which in themselves are not listed but are classified as toxic, the site becomes subject to CIMAH.

The proposed Control of Major Accident Hazards and Regulations[15] (COMAH) will incorporate the requirements of an amended directive (Seveso II) which, whilst similar to Seveso I and following the same two-tiered format for duties, differs in a number of important ways in that it:

(a) Emphasises the importance of safety management systems.
(b) Allows the directive to keep up to date with technical progress.
(c) Provides more details aimed at ensuring a more uniform implementation by Member States.

The main new features of this directive include:

(i) Dependence on the presence on site of threshold quantities of dangerous substances.
(ii) Widened scope to include explosives and chemical hazards at nuclear installations.
(iii) Greater use of generic categories of substances, i.e. toxic.
(iv) Introduction of a new ecotoxic category.

(v) Safety reports to be set out more precisely and made available to the public.
(vi) Land use planning requirements are to be introduced to ensure environmental risks are assessed.

7.6.3 The Chemicals (Hazard Information and Packaging for Supply) Regulations 1994 (CHIP 2) as amended in 1996 (CHIP 96) and the Carriage of Dangerous Goods by Road Regulations 1996 (CDG) and associated regulations

These major pieces of legislation replace the Classification, Packaging and Labelling of Dangerous Substances Regulations 1984 with the object of increasing the protection of people and the environment from the ill effects of chemicals. The Regulations do this by requiring the suppliers to:

(a) Identify the hazards of the chemicals they supply.
(b) Give information about the hazards to the people who are supplied.
(c) Package the chemicals safely.
(d) Label the packaged chemicals to identify the contents.

These are known as the *supply requirements*. A supplier is someone who supplies chemicals as part of a transaction and includes manufacturers, importers and distributors. Similar duties, known as the *carriage requirements*, are imposed on those who consign chemicals for transport by road. The transfer of chemicals between sites, even if under the same ownership, come within the scope of the Regulations which define the term 'chemical' to include pure chemicals, such as ethanol, as well as preparations and mixtures of chemicals, such as paints and pharmaceutical compounds.

The full package of legislation and supporting guidance for CHIP consist of:

(i) the Regulations[3]
(ii) the Approved Supply List[4]
(iii) the Approved Carriage List[20] (Note: The Approved Carriage List is not part of CHIP but of the Carriage of Dangerous Goods (Classification, Packaging and Labelling) and Use of Transportable Pressure Receptacles Regulations 1996 (CDG-CPL) but is included here for completeness.
(iv) approved guide to the classification and labelling of substances dangerous for supply[21] – CHIP 2
(v) approved code of practice on safety data sheets[5].

Wherever dangerous chemicals are supplied for use at work, CHIP 2 requires that safety data sheets must be provided when the chemical is first ordered and that they must contain sufficient information to enable the recipient to take the precautions necessary to protect his employees' health. The approved code of practice gives detailed advice on the information to be made available to employers. This could be useful when carrying out a COSHH assessment, but the data sheet should not be considered as a substitute for an assessment.

Dangerous chemicals supplied in a package must carry a label giving information about its hazards and the precautions to be taken in its use or in the event of a spillage. This label information should supplement the information given to employees in their training. For others, such as the emergency services, it allows them to take the correct precautionary action in the event of an incident.

The label on the package must contain details of:

(a) the supplier
(b) the chemical name
(c) the category of danger
(d) the appropriate risk and safety phrases.

Risk phrases summarise the main danger of the chemical, e.g. 'R45 – may cause cancer' or 'R23 – toxic by inhalation'. Safety phrases tell the user what to do, or what not to do, e.g. 'S2 – keep out of reach of children' or 'S 29 – do not empty into drains'. A warning symbol is also required on most labels, e.g. a skull and cross bones or a picture of an explosion. Suppliers are responsible for ensuring that the label carries the correct information. CDG-CPL also requires labels to be provided that warn those who handle packages of chemicals during transit on a public road. These transit labels are similar to, but not as comprehensive as, the packaging labels since they concentrate on the immediate information needed should the vehicle carrying them be involved in an accident. Where a package contains only a single substance, the requirements of both the supply and the carriage regulations may be combined into a composite label. An example of a label is shown in *Figure 7.3*.

When one or more receptacles are contained in a common outer packaging, the labelling may be in accordance with either of the two regulations (see section 7.8). Finally CHIP 2 requires chemicals to be packaged safely and in a manner to withstand the foreseeable conditions of supply and carriage.

In addition, NONS require manufacturers and importers to carry out a certain amount of testing before a quantity of new substance is placed on the market:

- 10 kg to 100 kg per year – very limited testing required
- 100 kg to 1 tonne per year – limited testing
- > 1 tonne per year – full testing required and a technical dossier to be produced.

When > 1 tonne of product is to be put on the market, the appropriate authority (Department of the Environment and HSE – and in Northern Ireland, the appointed authority) must be notified. The information to be included in the technical dossier is listed in schedule 2 to the Regulations and includes a hazard classification.

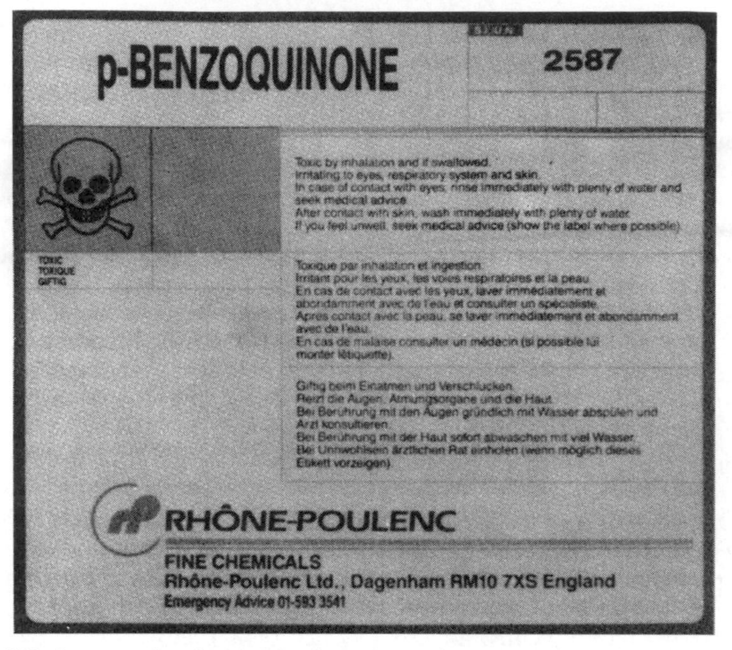

Figure 7.3 An example of a label for hazardous substance package showing warning sign. (Courtesy Ciba-Geigy Plastics)

7.6.4 The Control of Substances Hazardous to Health Regulations 1994 (COSHH)

These Regulations provide for a complete package of duties on health issues relating to the use, handling and storage of hazardous substances at work. In this, they place onerous duties on the employer as well as obligations on the employee.

Substances hazardous to health are defined in reg. 2 as any substance which:

- is listed in the approved supply list[4] as being very toxic, toxic, harmful, corrosive or irritant
- has been assigned an MEL or OES
- is a biological agent
- is dust of any kind in substantial concentrations in air
- any other substance creating like hazards.

Under these Regulations employers are required to:

1. Establish which of the substances they use are hazardous; what those hazards are and, if a new substance is proposed, what hazards it may present.
2. Carry out an assessment of the risks to their employees who may be exposed to hazardous substances.

3 Before a substance is used, assess the potential risk to their employees.
4 Design and install control measures to prevent or control, as far as is reasonably practicable, the exposure of employees to hazardous substances, except for carcinogens and biological agents where it is recognised that some exposure may not be preventable so a strict hierarchy of measures to be taken is laid down.

   Employees must not be exposed to levels above the MEL, but if this occurs the employer must investigate and take corrective action to prevent a repetition.
5 Ensure that the control measurers provided are properly used and looked after. Where these measures include plant or equipment, it must be properly maintained, regularly examined and tested and records kept of such tests and examinations.
6 Where there is a risk that employees are liable to be exposed to dangerous concentrations of a substance, that exposure must be monitored, recorded and the records kept for between 5 and 40 years, depending on the circumstances of exposure.
7 Where exposure to substances listed in schedule 5 occurs, health surveillance must be provided and medical records kept for at least 40 years.
8 Ensure that employees are:
   (a) Given information about the substances being used;
   (b) Made aware of the hazards those substances present;
   (c) Instructed in the procedure and techniques necessary for the safe use of those substances and any precautions that should be taken;
   (d) Provided with information on the results of monitoring and health surveillance; and
   (e) Given access to the COSHH Risk Assessment and any Occupational Hygiene Reports issued.

A major feature of COSHH is the duty to carry out assessments of the health risks associated with hazardous substances. There is no universally accepted approach to assessing risks and each company should do what is appropriate in the circumstances of their particular operations. An example of an assessment form is shown in *Figure 7.1(b)*.

The assessment should be carried out by someone who is familiar with the workplace, the process and the substances being used and who has sufficient knowledge to be able to interpret the implications of what is found. With the more complex processes, it has been found helpful to have a team carrying out these COSHH assessments; the team consisting of safety adviser with representatives from the chemists, production staff, engineers, operators and safety representatives. The assessment should consider the risks from not only those chemicals that are brought in to be stored, used or worked on in the workplace, but also by-products, intermediates and substances that are given off during the process or work activity as well as the finished product, residues, waste, scrap etc.

As part of an assessment, it is necessary to discover where and in what circumstances the substances are stored, used, handled, generated, released, etc., also what people do when handling the substances, as opposed to relying on the assumption that all work is carried out according to instructions. When carrying out a COSHH assessment the whole operation should be reviewed, i.e. from the preparation of the area to start the tasks, through handling the substances, to cleaning the area, equipment or plant. During the assessment, consideration should also be given to non-standard events that are likely to occur. The workforce should be able to provide historical evidence to help with this part of the review. The effectiveness of the measures taken to control and minimise exposures should be monitored since these are part of the overall system of work. The assessors should have a knowledge of the effect that the various substances have on the body and of the routes of entry, such as inhalation, ingestion, skin absorption, etc. Measured exposure levels should be compared with published exposure limits. The number of people exposed and the duration of exposure should also be noted.

The assessment should be recorded to document the conditions found – both good and bad – and the actions that were judged to be necessary to reduce any risk. If conditions change within the plant or process, the COSHH assessment should be reviewed and may need to be repeated. Advice and guidance on carrying out risk assessments are contained in HSE publications[22–24].

In endeavouring to meet the obligations posed by these Regulations, it is helpful to break the work into manageable parts, remembering the less visible people such as cleaners, storekeepers, and maintenance staff who can be exposed to substances often without check or control. A risk assessment should be carried out before any new maintenance activity is undertaken to ensure that correct control measures, training and system of work are agreed and followed.

7.6.5 The Environmental Protection Act 1990 (EPA)

This piece of legislation is the most profound development in British environmental law this century and contains many fundamental changes to pollution control regimes that are having far-reaching effects on industry. The Act is split into nine parts with Parts 1, 2 and 3 having the largest implications for industry. It is an enabling Act that specifies broad environmental objectives to be achieved and relies on subsidiary regulations to provide the details of the requirements.

7.6.5.1 Part 1: Integrated pollution control and air pollution control

This part deals largely with the administration of the Act outlining the authorisation of the bodies for enforcing it and the powers vested in them. It considers enforcement in two parts:

(i) Integrated pollution control (IPC) by the Environmental Agency (EA).
(ii) Air pollution control (APC) by local authorities (LAs).

The Environmental Protection (Prescribed Processes and Substances) Regulations 1991 allocate the enforcement of standards between the EA and the LAs.

The EA was formed under the Environment Act (1996)[32] by combining Her Majesty's Inspectorate of Pollution (HMIP), the National Rivers Authority (NRA), and the Waste Disposal Authority (WDA). The EA enforces the larger, more complex and polluting industries, e.g. power stations, chemical industry and heavy metal manufacture. Such processes are known as the Part A Processes and come under the IPC legislation. IPC deals with the environmental aspects of a process at all stages, aiming to achieve the least environmentally damaging combination of options, hence minimising pollution, and requires that the end disposal option must not be damaging to the environment. For example, an air pollutant may be removed by scrubbing the gas stream with water. However, what effect does the substance in water have on the sewage works or the end disposal source, the river? If the effect is detrimental then another disposal route must be chosen, e.g. thermal oxidation of the air stream. *Figure 7.4* shows a modern scrubbing plant for cleaning vapour discharges.

The conditions in an IPC authorisation cover emissions to water, land and air and extend to the prevention of persistent offensive odours at or beyond the boundary of the premises. The standards demanded by the authorisation will require the application of the *best available techniques not entailing excessive costs* (BATNEEC). This is a new enforcement concept seen as equivalent to *so far as is reasonably practicable* of HSW. The Environment Agency has produced numerous guidance notes on Part A Processes.

By contrast, APC deals only with emissions into the atmosphere, although LAs, who enforce it, are required to seek and abide by the decisions of the HSE on matters affecting human health, and of the EA where discharges to water courses are involved. This is a more limited approach than IPC and relates to processes having a supposedly lower pollution potential. Such processes are known as Part B Processes.

Employers are required to seek authorisation for processes involving discharges, whether to river, land or atmosphere, and should not discharge until an authorisation has been granted. However, where delays occur in granting authorisation, the processes concerned may be run but must meet current environmental standards. When granted, the authorisation will state the discharge standards to be achieved by the process. Failure to meet that standard may attract prosecution with, on conviction, penalties similar to those under HSW.

EPA empowers inspectors to issue Enforcement and Prohibition Notices which must specify the remedial steps to be taken. There is an appeal procedure against such notices to the Secretary of State who will appoint an arbiter.

Included in this part of the Act is a power to recover the cost of making the authorisations and of enforcement – on the principle of 'the polluter pays'. Fees will be charged for initial applications and for the annual renewal of the authorisations.

When the EA monitor discharges, they compare their findings with the appropriate authorised discharge limits and take action accordingly. They

Figure 7.4 Modern vapour-scrubbing plants to ensure compliance with atmospheric emission standards. (Courtesy Rhone-Poulenc-Rorer)

may employ specialist consultants to monitor the concentrations of air emissions from Part A Processes.

7.6.5.2 Part 2: Waste disposal[25–27]

Waste is any substance which is surplus to the process or company's requirements and can be a scrap material, an effluent, other unwanted substance or any substance or article that needs to be disposed of as being broken, worn out, contaminated or otherwise spoiled. Under the Special Wastes Regulations 1996 certain substances, listed in Schedule 1, have been classified as *special wastes* and their disposal must follow a strict laid down procedure.

A novel feature of EPA is the introduction of the concept of a 'duty of care' which places on all involved in the production and handling of waste a number of duties which cover five aspects:

1 To prevent the keeping, treatment or disposal of waste without a licence or in breach of a licence.
2 To prevent the escape of waste.
3 To transfer waste only to an authorised person.
4 To ensure that there is clear information and labelling of the waste.
5 To retain documentary evidence.

The major responsibility for ensuring that these duties are fulfilled rests with the creator of the waste. Thus, the producers of waste have an obligation to prevent the escape of waste along the whole length of the disposal chain through the use of reputable disposal companies and registered hauliers. Transporters of waste have to be registered with the county council for the area in which they are based. Producers must ensure that waste is carried by authorised carriers and disposed of at facilities which have planning permission, are licensed and that the licence allows that type of waste to be deposited. Producers must also satisfy themselves that the waste can be stored securely without loss of containment and that the method of disposal is appropriate.

There must be a tight control on the documentation for the movement of the waste to demonstrate fulfilment of the duty of care. Producers must periodically audit each stage of the waste disposal operation. The penalties for non-compliance are similar to those under HSW.

The standards for the keeping and disposal of waste are being significantly tightened with land-fill licence conditions becoming more stringent, usually by requiring better leachate control, methane monitoring, aquifer monitoring, etc. More frequent inspections will be carried out by inspectors of the Environment Agency which has taken over the waste regulation function previously carried out by the county councils. As with IPC, recovery of the cost of enforcement and administration of waste control is introduced. Regulations may be made to extend the definition of 'special wastes' to encompass many of the known hazardous substances in current use. Other regulations may provide for a list of exemptions from licensing for substances even though they are still classified as special waste.

7.6.5.3 Part 3: Statutory nuisances

A statutory nuisance is the emission of smoke, fumes, gases, dust, steam, smells, other effluvia and noise at a level which is judged to be prejudicial to health or a nuisance to the community or anyone living in it. The legal interpretation is somewhat flexible.

Where a local authority is satisfied that either a statutory nuisance exists or is likely to occur or recur, it can serve an Abatement Notice requiring the abatement, prohibition or restriction of the occurrence or recurrence. The notice will specify the date by which the notice is to be complied with and may also specify the remedial action to be taken. Appeals against the notice can be made within 21 days. Failure to comply with the notice can attract a fine on conviction of up to £20 000. It is a defence to a prosecution to demonstrate that the best practicable means had been used to prevent or counteract the effects of the nuisance. A private individual can start a proceeding in a magistrates' court and, if the court is satisfied that a nuisance exists, it can require the accused to abate or terminate it. The court may also specify the remedy to be followed.

7.6.5.4 Summary

Governmental and public pressure is demanding a significant tightening of environmental standards. Much of the onus for achieving this will fall on industry which may be involved in considerable expense in carrying out the work necessary to meet the new standards.

7.7 Storage of substances

The HSE has issued a number of guidance notes on the storage of specific substances such as chlorine[28], LPG[29], highly flammable liquid[30] etc. Further guidance can be obtained from the distributive trade association[31] or from the supplier of the substance who will offer practical advice on standards of storage of his products. Some suppliers provide a 'Duty of Care Service' by visiting sites to assess the bulk storage conditions of the materials they supply.

An overriding principle in the storage of chemicals is that they should not be adversely affected by other adjacent substances or operations. An HSE Guidance Note gives suitable information on this topic[32]. The proposed COMAH Regulations[15] also require consideration to be given to the quantities of hazardous materials that are stored at chemical sites located near to a COMAH site because of the possible 'domino effect' of an incident at one company affecting adjacent companies.

7.7.1 Drum compounds and storage tanks

A common problem met in factories using chemicals is where to store flammable solvents, or substances that could give off a hazardous vapour if a leak occurred. Whether in drums or bulk tanks, outside storage is preferable since this:

(a) Allows dispersion of any flammable or hazardous vapour from vents, leaks, spillages etc.
(b) Minimises the potential of ignition.
(c) Provides secure storage to minimise damage to containers.

External storage of chemicals and hazardous substances require particular facilities which, for bulk tanks, drums and intermediate bulk containers (IBCs), include:

(a) a bunded impervious hard-standing (e.g. a concrete plinth) capable of containing any spillage or rainwater,
(b) drainage of the impervious bunded area to a sump, which can be pumped out for authorised disposal, dependent on the hazardous nature of the liquors,
(c) a separation of at least 4 metres from boundaries, building and sources of ignition,
(d) except for bulk tanks, a ramp to allow trolleys and fork-lift trucks easy access for handling the drums or IBCs,
(e) segregation of stocks of materials which if mixed, due to a spillage or incident, could cause a hazardous reaction,
(f) measures to ensure that containers of flammables are not stored under overhead electric power lines.

By ensuring that all drums and IBCs are stored in a bunded area, the risk of ground or surface water pollution is reduced. Such a facility also ensures the flammables are contained, reducing the risk of the spread of fire should a leak occur.

A responsible person should be put in control of the drum compound with duties that should include:

- regular housekeeping checks
- arranging for water in sumps to be pumped out for appropriate disposal
- arranging for spillages/leaks to be dealt with
- maintaining an inventory of materials stored on the compound.

In the case of bulk storage, the tank should be mounted above ground level and be provided with:

(a) Safe access and egress for delivery or waste disposal tankers.
(b) Adequate separation distances from other tanks and sources of ignition.
(c) Hazardous electrical zone classification when handling flammable materials.
(d) An impervious bund capable of containing 110% of the tank capacity.
(e) A tank contents gauge arranged for local or remote reading.

(f) A pressure/vacuum relief valve or vent to allow the tank to breathe during filling and emptying.
(g) Secure attachment to the ground to prevent movement due to high winds.
(h) Fixed installation for fire fighting to be considered, e.g. sprinklers, drench or foam.
(i) Adequate HAZCHEM labelling.

Figure 7.5 shows a bulk storage tank for non-flammable liquids.

Figure 7.5 Storage tank for non-flammable liquids. (Courtesy Rhône-Poulenc-Rorer)

Underground storage tanks (e.g. petrol stations) are required to be fitted with a means for detecting leaks[33].

The tanker off-loading bay – ideally sited off road – should be drained to a sump to prevent water pollution. Safety shower and eye wash facilities should be provided in the area. The location of the emergency equipment should not be within the confines of a potential leak/spray from the transfer hose.

7.7.2 Tanker off-loading

The tanker off-loading point should be clearly labelled and locked off to prevent unauthorised access. There have been a number of incidents involving tanker loads being off-loaded into the wrong storage tanks. It is important that a representative of the company owning the storage tank is present during the off-loading in order to:

- Check delivery notes and paperwork.
- Unlock the off-loading points access pipe.
- Ensure correct procedures and PPE are adhered to.

Where non-compatible substances are delivered to the same area of a site, the transfer hoses should be arranged to be unique to the particular substance, for example by using different threads on the hose fittings.

Accidents have occurred when sampling road tankers before they are off-loaded. Where this is necessary, the responsibility and the procedure for carrying out such an operation should be agreed in writing with the supplier of the material. The sampling procedure must clearly define the responsibilities of both the customer and the supplier, paying particular attention to:

- Personal protective equipment.
- Sampling equipment and method of taking samples.
- Safe system of work when working on top of the tanker. (Ideally, a walkway adjacent to, or on the top of, the tanker should be provided.)

7.7.3 Warehousing

The first step when chemicals and substances are to be stored is to know what the substances are and their characteristics. This will enable storage to be arranged so that incompatible substances are either suitably separated, have an intermediate fire wall or a mutually compatible substance between them to provide such separation. The effects of storing incompatible chemicals in the same area have been described in HSE reports[34,35].

Fire detection and sprinkler systems are beneficial but if used with high bay racking consideration should be given to incorporating inter-rack sprinklers which counteract the chimney effect of a fire. Care should be taken that no water-reactive chemicals are stored below sprinkler installations. The advice of the local fire prevention officers, sprinkler suppliers, insurance surveyors and safety advisers should be sought on storage and rack layouts and on the arrangement of the sprinklers.

Following an incident at Basle, Switzerland, in which contaminated fire-fighting water from a warehouse fire entered the Rhine and caused severe ecological damage, the chemical industry and storage companies have been studying better methods of run-off water control and their recommendations are contained in a booklet[31].

7.7.4 Storage of gas cylinders

Gases stored in cylinders under high pressure are used extensively in industry for burning, welding and as process material. The contents of the containing gas cylinders are potential sources of fire and explosion if the cylinders are not stored safely[36]. They should be stored upright and checked regularly, particularly the regulators. Gas cylinders should be stored away from lifts, stairs, gangways, underground rooms, and in an area that is free from fire risks and sources of heat and ignition. In order to prevent the bottom of a stored upright cylinder from corroding, cylinders should be stored under cover on a well-drained surface. Manufacturer's recommendations should also be consulted when storing different types of gases in the one area. This information will include segregation and separation distances for cylinders in storage.

Where gases such as nitrogen and carbon dioxide are used in a small enclosed room (e.g. a laboratory or switch room) any leakage can create an asphyxating atmosphere. Therefore, precautions need to be taken to ensure the atmosphere is checked before anyone enters the area.

7.8 Transport

Chemicals are transported either in discrete packages or in bulk, often by tanker. Careful selection must be made of the material and construction of the package or container: for example, a mild steel road tanker will not last long if used to transport 20% sulphuric acid, whereas it is perfectly satisfactory for carrying fuel oil. Where the substance is packaged in small quantities, the packaging and labelling must comply with the Carriage of Dangerous Goods by Road Regulations (1996). These Regulations are supported by Guidance Notes and Approved Codes of Practice[37,38].

> *Note*: The vehicles must carry a suitable warning sign. These Regulations and the Approved Carriage List[20] cover the suitability of the vehicle and container and the information to be made available about the substance(s) being carried. Regulations[39] provide information about the type and standard of instruction and training of drivers of vehicles carrying dangerous goods. Other Regulations[40] lay down the requirements for the safe transport of explosives by road.

Vehicles carrying more than 500 kilograms of dangerous substances should carry a hazard warning board at the front and rear of the units showing the following information:

- HAZCHEM code.
- Substance identification number.
- Hazard warning sign.
- Contact telephone number for technical advice.

The Regulations for transport in packages and in bulk have similar requirements for the proper stowage of the load, adequate instruction and training for drivers, information on the load being carried and proper labelling of the vehicles. A review of the safe handling of chemicals and dangerous substances in road transport operations is contained in a chemical industry publication[41].

The Carriage of Dangerous Goods by Rail Regulations (1996) and the Packaging, Labelling and Carriage of Radioactive Material by Rail Regulations (1996) cover the suitability of the rail freight/containers and information to be made available when transporting dangerous goods and radioactive material by rail.

The HSC are due to publish a Consultative Document[52] on 'Proposals for the Implementation of Dangerous Goods Safety Adviser Directive' which will contain a proposal that companies which transport dangerous goods should employ an expert in this branch of safety.

Consideration should be given to the safe transportation of substances on company premises:

1. If drums are carried long distances, they should be transported on pallets on a towed trailer.
2. Drum handling equipment should be regularly maintained.
3. Road surfaces should be free of obstacles that could cause a spillage or a load to become unstable.
4. Winchesters of liquid should be carried in protective containers fitted with a handle.

7.9 Plant and process design

The chemical manufacturing process should be operated and maintained in a safe manner. This presumes that the design of the plant is safe. To design a plant and operating process that will be safe requires knowledge of the chemical process and its limiting parameters. The design of the plant should consider not only matters concerned with safety, but also health, the environment, fire, and its effect on neighbours. Once the plant has been installed, regular checks should be carried out to ensure proper compliance with agreed standards. Such checks should include house-keeping, scheduled maintenance and health, safety and environmental audits. Changes to the plant should only be made after discussion and investigation of any likely effects the proposed changes may have on the plant safety.

When new processes are proposed, the safety adviser should be involved at the earliest stage and should ensure that adequate chemical data, environmental, fire and safety information are available. Discussions on the plant may involve the enforcing authorities and waste disposal agencies but their early involvement could save delays at a later date.

7.9.1 Process safety

This section relates to the safe design of process plant which starts with the chemist developing a chemical reaction or series of reactions that produce the required end product. Initially the proposed process will be tried out in the laboratory in gram quantities. When the feasibility of the process is proved a pilot scale plant will be built which will carry out the synthesis in kilogram quantities to establish the manufacturing viability of the process. At this scale of operation, the chemist will be looking not only for product purity and quality but also at some of the physical chemistry aspects such as heats of reaction, temperature rises, rates of pressure rise, temperature of first exotherm etc.

The information gathered from the pilot plant study is compiled into a process dossier which provides the data for the chemical engineer to use in the design of the full-scale production plant. It is at this stage that the safety adviser should become involved.

The initial design will be in the form of engineering line diagrams which show the vessels, pumps, interconnecting pipework, etc., but not the detailed instrumentation or control systems. Sufficient information should be available to enable a preliminary safety analysis to be carried out drawing on the expertise of the process chemist, process engineer, control systems engineer, production engineer and the safety adviser. The object of this safety analysis is to identify possible hazards arising from the proposed plant and to define the measures necessary to eliminate or reduce them to an acceptable level. Some of the more obvious hazards to be looked for will be low or high pressures, low or high temperatures and highly exothermic reactions which could cause a runaway reaction with potentially catastrophic results.

In the analysis, each process step is looked at for deviations in temperature, pressure, level etc., and the possible effect of physical problems such as pipe blockage, valve failure, corrosion etc. All the details of this analysis should be recorded. On the findings of this analysis the chemical engineer can refine his design and produce detailed plans including pipework layouts and instrumentation (P & I) diagrams.

Once the detailed plans have been prepared, a more detailed safety analysis can be carried out which could be in the form of a Hazard and Operability Study (HAZOP)[42,43].

7.9.2.1 Hazard and Operability study (HAZOP)

A HAZOP study is a technique used to identify health, safety, environmental and fire hazards and operating problems. The basic study deals with the mechanical, electrical and chemical aspects of the plant operation together with electromechanical control systems. Where the control equipment incorporates computers additional studies are needed.

A HAZOP study requires a multi-disciplinary approach by a team made up of technical specialists, i.e. chemical engineer, chemist, production manager, instrumentation engineer, safety adviser etc. It is co-ordinated by a leader who guides the systematic investigation into the effect of various faults that could occur. The success of this study relies heavily on the quality of the leader and the positive and constructive attitude of the team members. It is essential that the team have all the basic data plus line diagrams, flow charts etc., and understand how a HAZOP study works.

The HAZOP study breaks the flow diagram down into a series of discrete units. Various failure and fault conditions are then considered using a series of 'guide words' to structure the investigation of the various circumstances that could give rise to those faults. Each deviation for each guide word is considered in detail and team members are encouraged to think laterally and to ask questions especially about the potential for causing a fault condition. *Table 7.1* shows how each of the guide words can be interpreted to highlight possible deviations from normal operation and *Figure 7.6* shows a HAZOP report form that could be used to record the findings of the study.

In the example in *Table 7.1*, under the first guide word, 'None', we could ask:

- What could cause no flow?
- How could the situation arise?
- What are the consequences of the no-flow situation?

Table 7.1. Showing typical interpretations of HAZOP guide words

Guide word	*Deviations*
None	No forward flow, no flow, reverse flow.
More of	Higher flow than design, higher temperature, pressure or viscosity etc.
Less of	Lower flow than design, lower temperature, pressure or viscosity etc.
Part of	Change in composition, change in ratio of components, component missing.
More than	More components present in the system, extra phase, impurities present (air, water, solids, corrosion products).
Other than	What else can happen that is not part of the normal reaction, start-up or shutdown problems, maintenance concerns, catalyst change etc.

HAZARD AND OPERABILITY STUDY

Date of Study / / Sheet of sheets

Study title ______________ Study team ______________

Prepared by ______________ Project number ______________

Line diagram number ______________ Procedure number ______________

Step Number/ Guide Word	Deviation	Cause	Consequence	Action

Figure 7.6 Report from a HAZOP Study

- Are the consequences identified hazardous or do they prevent efficient operation?
- If so, can we prevent no-flow (or protect against the consequences) by changing the design or method of operation.
- If so, does the size of the hazard (i.e. severity of the consequences multiplied by the probability of the occurrence) justify the extra expense?

Similar questions are applied to each the other guide words, and so on. Each time a component is studied the drawing or diagram should be marked. Not until all components have been studied can the HAZOP study be considered complete. Where errors occur on the drawing or more informations needed the drawing should be marked (using a different colour) and the points noted in the report.

To be effective the team needs to think laterally and there should be no criticism of other team members' questions. A strange or oblique question may spark off a train of investigation which could lead to the identification of potentially serious fault conditions.

A well-conducted HAZOP study should eliminate 80–85% of the major hazards, thereby reducing the level of risk in the plant. In safety critical plant another HAZOP study, carried out when the detailed design has been finalised, could increase the probability of safe operations.

When the HAZOP study has been completed the necessary remedial actions should be agreed for implementation by the project or process manager. Records of the changes in the design should be kept and checks made to ensure that the modifications have been carried out during the construction of the plant.

With plant that is controlled by computer, the HAZOP study needs to include consideration of the effects of aberrant computer behaviour and the team carrying out the study may need to be reinforced by the software designer plus an independent software engineer able to question the philosophy of the installed software program. A technique known as CHAZOP has been developed for such plant which also highlights the safety critical control items.

7.9.3 Plant control systems

Many small, simple, and relatively low hazard plants are fully manually operated. However, with more complex plant automated controls using electronic control systems are employed This does not necessarily make it safe since faults can, unknowingly, be built into the controlling software. To achieve optimum levels of safe operation, computer software for plant control systems should be devised jointly by the software specialist and the production staff. All operational requirements must be covered to ensure that the software designer does not make assumptions which could result in faulty or even dangerous operation of the plant. The software must be designed to accommodate plant failures and any testing or checking necessary during or following maintenance.

Before installing it, the computer program must be challenged in all possible situations to ensure that it matches operational requirements. Any review of software should include an independent software engineer who can challenge the philosophy behind the software. All software changes must be fully described and recorded and plant operators fully trained in the effects of the changes.

When automated computer control systems are incorporated into a plant, operators tend to rely on them completely to the extent that there is a risk that they forget how to control the plant manually. This can be critical in an emergency and it may be prudent to switch the computer off occasionally, and, under supervised conditions, ensure that the operators are still able to control the plant manually.

Control panels should not be provided with too many instruments since this can confuse the operator and prove counterproductive. However, sufficient instrumentation is needed to enable the operators to know what is going on inside closed vessels, pipes, pumps etc. Critical alarms should be set into separate parts of control panel to highlight their importance. This will reduce the potential for their being confused with others, and possibly overlooked. The tone of audible critical alarms should be different from that of process alarm systems to prevent confusion.

Computer-controlled plant will frequently have three levels of operational and safety control:

Level 1: Will mostly focus on process control of the plant and give indicative warnings of possible safety concerns when, for example, a rapid temperature rise may trigger a warning panel indicator.

Level 2: Control occurs when computer software initiates changes to control reaction kinetics. If a reaction temperature continues to rise, the software would initiate the application of cooling water to the vessel to regain control and continue production.

Level 3: Is entirely a safety system when the process is out of control. It will rely on hard-wired trips that shut the plant down safely and abandon production. The hard-wired trips work independently of the computer system.

There is no universal formula for control systems and a control strategy must be developed for each plant based on the operating parameters. A small batch plant consisting of two chemical reactors having a mixture of manual and automatic controls is shown in *Figure 7.7*.

7.9.4 Assessment of risk in existing plants

A review of existing chemical facilities should be undertaken to identify possible faults and so avoid acute and/or catastrophic loss. The assessment should focus on 'instantaneous failure prevention' of plant such as:

Figure 7.7 Two chemical reactors having manual and automatic controls. (Courtesy Rhône-Poulenc-Rorer)

- bulk oil or chemical storage facilities
- multi-chemical 200 litre drum store (especially if large-scale dispensing is carried out)
- chemical processes or mixing facilities
- solvent recovery plant
- pipelines and pipework that contain oils or chemicals in quantity and/or under pressure.

This review will identify those systems or processes that require a further detailed study which can be carried out using one or more of the techniques described below. The end result of the assessment should be a position statement which describes the level of risk from the plant and identifies which facilities require additional measures to ensure they remain both physically and environmentally safe.

A number of techniques have been developed to identify the hazards and to assess the risks from plant and equipment. These techniques range from the relatively simple to the highly complex. A number are described in a BS EN standard[44]. Whichever technique is used it should be appropriate to the complexity of the plant and the materials involved.

7.9.4.1 Simpler techniques

The simpler techniques are aimed primarily at determining a ranking order of the risks from the chemical processes carried out in the area. They should clarify which facilities create insignificant risks and require no further action. The position statement for these facilities should record the reasons for this decision. The simpler techniques include:

1. The 'What-if method' is the simplest method to assess chemical process safety risks and is based on questions such as 'What if the mechanical or electrical integrity of the process, the control systems and work procedures all fail, . . . what consequences could arise in the worst case?' While the potential consequences are largely determined by the inherent hazard of the material and the quantity involved, the reviewer is focused on safety concerns, e.g. those arising from fire, explosion, toxic gas release, and environmental protection.
2. The 'Checklist method' is a structured approach whereby the reviewer responds to a predetermined list of questions. This method is less flexible than the 'What-if method' and its effectiveness relies on the strengths and weaknesses of a predetermined checklist. Examples of checklists can be found in chemical process safety literature.
3. The 'Dow-Mond Index' is a more structured approach than the previous two techniques and takes into account quantities and hazards to arrive at a basic risk classification. This method provides a level of quantification of risk and considers the 'off-setting' factors which exist to control intrinsic hazards.

7.9.4.2 More complex techniques

Where the ranking process, described above, identifies facilities that warrant an assessment in greater depth, one of the techniques described below should be used:

1 HAZOP study (see section 7.9.2.1).
2 Failure modes and effects analysis (FMEA). FMEA is an inductive method for evaluating the frequency and consequence of failures. It involves examining every component and considering all types of failure for each. It can indicate generic components that may have a propensity to fail.
3 Fault tree analysis (FTA)[45] FTA is a deductive method which starts by considering a particular fault or 'top event' and works backwards to form a tree of all the events and circumstances that could lead to the happening of that top event. By assessing the probability of each individual event, an estimate of the probability of the top event occurring can be obtained. If that probability is unacceptable the major components contributing to it can easily be identified and a cost-effective replacement of them implemented. This method lends itself to assessing the impact of changes in the system and has been useful in determining the causes of accidents.

7.9.5 Functional safety life cycle management (FSLCM)[46]

FSLCM is a new technique designed to enable plant safety systems to be managed in a structured way. The technique has been designed to accommodate computer-controlled plants from start-up to shutdown, including emergency shutdowns. It aims to ensure that the safety related systems which protect and control equipment and plant are specified, engineered and operated to standards appropriate to the risks involved. The key concepts of this technique are:

(a) *The safety life cycle* – begins with a clear definition of the equipment and processes for which functional safety is sought and by a series of phases provides a logical path through commissioning, operation to final decommissioning.
(b) *Safety management* – sets a checklist for the things that need to be in place in order to prepare for and manage each phase of the *safety life cycle*. These are incorporated into a formal safety plan.
(c) *Design of safety related systems* – puts the design of safety related control and protective systems into the overall context of the safe operation of equipment or facilities. It requires that such systems are designed to meet specific risk criteria.
(d) *Competencies* – provides guidance on the appropriate skills and knowledge required by those people who will be involved in the technique.

By following a structured life cycle approach the hazards inherent in the operation of equipment or processes can be clearly identified. The standards to which protection is provided can be demonstrated in an objective and constructive way.

7.10 Further safety studies

Having carried out a HAZOP study on the plant and incorporated its findings into the design, it is prudent to carry out a further review during the commissioning period to check that the design modifications have produced the desired results. This is necessary since the final details of the physical installation are often left to the installing engineers to decide and these could produce unforeseen hazards. Finally, once the plant is commissioned and operational there should be routine safety checks carried out on a regular basis.

7.11 Plant modifications

Plant modifications, even apparently simple ones, can have major consequential effects[14]. It is crucial that the plant is not modified without proper authorisation and, for safety critical parts, the completion of a HAZOP study of the possible effects of the proposed changes. A 'process change form' should be used which should include the reasons for the change. Use of such a form also ensures a degree of control on the modifications made, especially if it has to be sanctioned by a senior technical specialist such as a process engineer, safety adviser, production manager and maintenance manager. There needs to be clear guidance as to when the process change form has to be used so that there can be no misunderstanding. After the plant has been modified it may be necessary to retrain the operators in the changed operation techniques.

7.12 Safe systems of work

Since human beings are necessary in the operation of chemical plants there is always the likelihood of errors being made that could result in hazards. It is, therefore, important that operators are trained in the safe way to run the plant. Such training, based on safe systems of work, should include the carrying out of risk assessments. Errors in operation and misunderstandings can be reduced if the system of work is in writing.

7.12.1 Instruction documentation

There should be detailed written operating instructions for every chemical plant which can conveniently be considered in three parts:

1 *Operator's instructions* that give specific instructions on how to operate the plant and handle the materials. They should contain information on

the process, quantities and types of materials together with any special instructions for dealing with spillages or leaks, temperatures and pressures to be expected and action to be taken if they are exceeded, first aid and emergency procedures, PPE to be worn, a copy of the safety data sheet for each of the materials involved, techniques for taking samples and cleaning instructions.

2 *Manufacturing procedures* aimed at the supervisor, chemist or engineer in charge of the plant should explain the process with a synopsis of the chemistry and refer to likely problems such as exotherms, give the sequence of operations, quantities of materials, temperature and pressure ranges, methods for dealing with spillages and leaks, disposal of waste etc.

3 *Process dossier* has been referred to above and should contain detailed information about the process and the plant. It would be a major reference for the process engineer.

7.12.2 Training

Both operators and supervision should be trained in the techniques for operating the plant, the process, materials used, their hazards and precautions to be taken, emergency procedures and first aid. The training can be based on the content of the Operator Instructions and the Manufacturing Procedures and should include a study of the safety data sheets. The importance of following the safe methods of work and the reporting of any deviations from the stated operating parameters should be emphasised.

7.12.3 Permits-to-work

Permits-to-work are required where the work to be carried out is sufficiently hazardous to demand strict control over both access and the work itself. This can occur when maintenance and non-routine work is being carried out in a chemical plant or for any normal operation where the risks faced make clear and unequivocal instructions necessary for the safety of the operators.

The essential elements of a permit to work are:

(a) The work to be carried out is described in detail and understood by both the operators of the plant and those carrying out the work.
(b) A full explanation is given to those carrying out the work of the hazards involved and the precautions to be taken.
(c) The area in which the work is to be carried out is clearly identified, made as safe as possible and any residual hazards highlighted.
(d) A competent, responsible and authorised person should specify the safety measures, such as electrical isolation, pipes blanked off etc. to be taken on the plant, check that they have been implemented and sign a document confirming this and that it is safe for workmen to enter the area.

(e) The individual workmen or supervisor in charge must sign the permit to say they fully understand the work to be done, restrictions on access, the hazards involved and the precautions to be taken.
(f) The permit must specify any monitoring to be carried out before, during and after the work and require the recording of the results.
(g) When the work is complete, the workmen or supervisor must sign the permit to confirm that the work is complete and it is safe to return the plant to operations.
(h) A competent, responsible and authorised person must sign the permit, cancelling it and releasing the plant back to operations.

The format of a permit to work will be determined by the type of work involved but a typical permit is shown in *Figure 7.8*.

Typical work requiring a permit to work includes hot work, entry into confined spaces, excavations, high voltage electrical work, work involving toxic and hazardous chemicals etc. For a permit to work to be effective it is essential that all those involved understand the system, the procedure and the importance of following the laid down procedure. Before the work starts all those concerned should be trained in the system and their individual responsibilities emphasised.

7.13 Laboratories

The use of chemicals in laboratories poses totally different problems from those met in a production facility. The scale is much smaller, the equipment generally more fragile and, while the standard of containment for bench work is often less, the skill and knowledge of those performing the reactions are very high.

Work in quality control laboratories is normally repetitive using closely defined analytical methods. Research laboratories are far wider in the scope of the reactions they investigate, sometimes dealing with unknown hazards, and in the equipment they use. The principal hazards met in laboratories are fire, explosion, corrosion, and toxic attacks. A limit should be specified for the total amount of flammables allowed in a laboratory at any one time, which should be enough for the day's work but not exceed 50 litres.

Hazardous and potentially hazardous reactions should be carried out in a fume cupboard. The effectiveness of the fume cupboard's extraction should be checked regularly in line with COSHH requirements. The fume cupboard should not be used for extra storage space since this can reduce the efficiency of the extraction system. A well-ordered and tidy fume cupboard is shown in *Figure 7.9*.

Further measures that can improve laboratory safety include:

(a) Instituting a 'peer review' assessment by asking a competent colleague to review the proposed reaction before allowing experiments to be carried out.
(b) Regular checks of laboratory storage areas to ensure old stocks and out-of-date reactive chemicals (e.g. peroxides formers) are removed for disposal. Holding minimum inventories of chemicals.

X Y Z Company Limited

PERMIT-TO-WORK

NOTES:

1 Parts 1, 2 and 3 of this Permit to be completed before any work covered by this permit commences and the other parts are to be completed in sequence as the work progresses.
2 Each part must be signed by an Authorized Person who accepts responsibility for ensuring that the work can be carried out safely.
3 None of the work covered by this Permit may be undertaken until written authority that it is safe to do so has been issued.
4 The plant/equipment covered by this Permit may not be returned to production until the Cancellation section (part 5) has been signed authorizing its release.

PART 1 DESCRIPTION

(a) Equipment or plant involved ______________________

____

(b) Location ______________________

(c) Details of work required ______________________

____

Signed ______________________ Date ______________
person requesting work

PART 2 SAFETY MEASURES

I hereby declare that the following steps have been taken to render the above equipment/plant safe to work on: ______________________

____

Further, I recommend that as the work is carried out the following precautions are taken: ______

____

Signed ______________________ Date ______________
being an authorized person

PART 3 RECEIPT

I hereby declare that I accept responsibility for carrying out the work on the equipment/plant described in this Permit-to-Work and will ensure that the operatives under my charge carry out only the work detailed.

Signed ______________ Time __________ Date __________

Note: After signing it, this Permit-to-Work must be retained by the person in charge of the work until the work is either completed or suspended and the Clearance section (Part 4) signed.

PART 4 CLEARANCE

I hereby declare that the work for which this Permit was issued is now completed/suspended* and that all those under my charge have been withdrawn and warned that it is no longer safe to work on the equipment/plant and that all tools, gear, earthing connections are clear.

Signed ______________ Time __________ Date __________
* delete word not applicable

PART 5 CANCELLATION

This Permit-to-Work is hereby cancelled

Signed ______________ Time __________ Date __________
being a person authorized to cancel a Permit-to Work

Figure 7.8 Permit to work

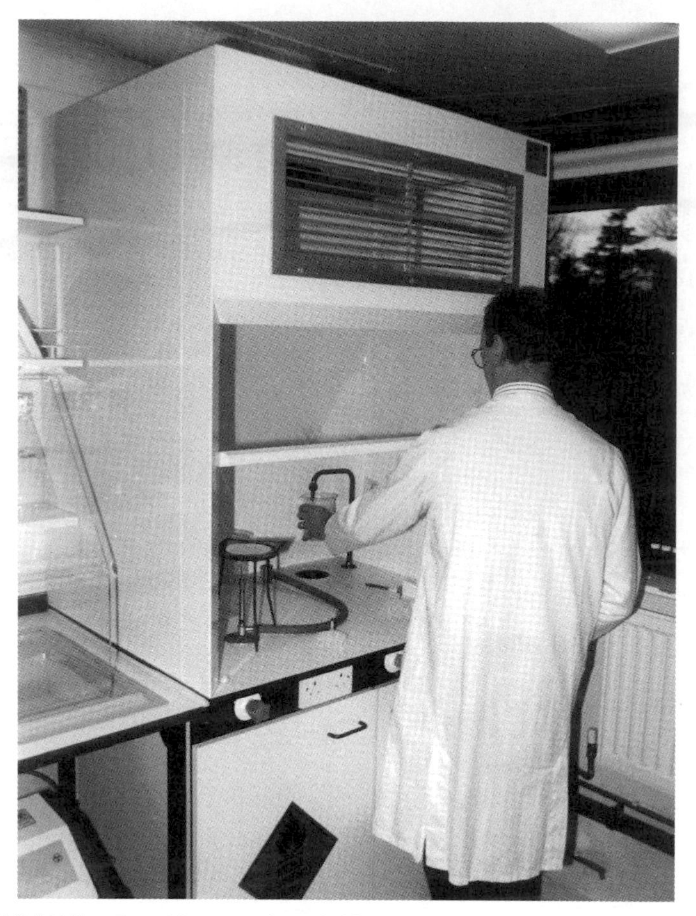

Figure 7.9 Well ordered fume cupboard. (Courtesy British Sugar plc)

(c) Producing a laboratory safety manual and regularly training staff in its contents.
(d) Providing spillage cleaning equipment and adequate training in its use.
(e) Establishing safe waste disposal procedures.
(f) Maintaining a high standard of housekeeping.

Laboratory safety is a very wide subject and there are a number of publications giving sound guidance[47–49]. Many of the larger chemical manufacuring companies produce their own practical guidance and are pleased to supply copies.

7.14 Emergency procedures

The Management of Health and Safety at Work Regulations 1992 (MHSWR) imposes on employers an explicit duty to have in place effective procedures to be followed in the event of serious or imminent danger to people at work. The CIMAH Regulations also require affected manufacturers to prepare on-site emergency plans. In addition, CIMAH requires employers to co-operate with the local authority in developing off-site emergency plans. Irrespective of these statutory requirements it is prudent for every user and storer of hazardous substances to prepare an emergency plan to cover all reasonably foreseeable events such as fire, major spillage or toxic release. The plans can be at two levels, one for the immediate production or storage area and the second for the site as a whole taking account of the likely effects on the local community.

It is very important that employees and the local emergency services know exactly and unambiguously what to do should an incident occur. The Dangerous Substances (Notification and Marking of Sites) Regulations 1990 require that the entrances to sites are labelled such that the emergency services have pre-warning that there are hazardous chemicals on site. Additionally, the Planning (Hazardous Substances) Regulations 1992 require notification to the local authority of the amounts of hazardous substances held on site. A clear drawing or sketch showing the layout of the site should be available for the emergency services. It should also contain details of the buildings and highlight fire extinguishers, emergency exits, spillage control equipment, etc. All employees should be properly instructed, fully trained and rehearsed in those emergency plans. The local emergency services should be encouraged to familiarise themselves with the site.

Where there is a potential for a major emergency, which would involve the local emergency services and local authority, there must be an agreed plan of action to co-ordinate all the services including managers and employees on the site with their specialised knowledge of the site and its processes. The emergency plans should include a list of emergency contacts including such bodies as the local authority, the water authority, the factory inspectorate, the pollution inspectorate, the police, etc.

Advice and guidance on preparing emergency plans are contained in publications by The Society of Industrial Emergency Services Officers (SIESO)[50] and the Chemical Industries Association (CIA)[51]. It must be emphasised that all emergency plans must be regularly practised and reviewed – there is no substitute for actually doing it!

7.15 Conclusions

This chapter has summarised some of the health, safety and environmental problems posed by the use of chemicals. A systematic review has been applied in an attempt to clarify the issues and facilitate an understanding of legislative requirements and good practices. Those with responsibilities for handling and using chemicals should study the relevant laws and guidance to ensure that their areas of responsibility meet the highest standards. Management commitment, leadership and setting a good example play important roles in achieving high standards in health, safety and the environment which, in turn, lead to a successful enterprise. To quote the HSC's slogan 'Good health is good business'.

References

1. *The Control of Substances Hazardous to Health Regulations 1994*, The Stationery Office, London (1994)
2. Health and Safety Executive, Legal Series Booklet No. L5, *General COSHH ACOP and Carcinogens ACOP and Biological Agents ACOP* (1996 edn), HSE Books, Sudbury (1997)
3. *The Chemicals (Hazard Information & Packaging for Supply) Regulations 1994*, The Stationery Office, London (1994)
4. Health and Safety Commission, Legal Series Booklet No. L76, *Approved Supply List. Information approved for the classification & labelling of substances and preparations dangerous for supply* (3rd edn), HSE Books, Sudbury (1997)
5. Health & Safety Executive, Legal Series Booklet No. L62, *Safety data sheets for substances and preparations dangerous for supply. Guidance on Regulation 6 of CHIP Regulations 1994. Approved Code of Practice*, HSE Books, Sudbury (1995)
6. *The Merck Index*, 10th edn, Merck & Co. Inc. (1983)
7. Sax, N.I., *Dangerous Properties of Industrial Materials* (7th edn), Van Nostrand Reinhold (1989)
8. Bretherick, L., *Handbook of Reactive Chemical Hazards*, Butterworth, Oxford (1979)
9. Health & Safety Executive, Guidance Series Booklet No. HS(G) 117, *Making Sense of NONS. A Guide to the Notification of New Substances Regulations 1993*, HSE Books, Sudbury (1994)
10. European Union, *European Inventory of Existing Commercial Substances*, EU, Luxembourg
11. European Union, *European List of Notified Chemical Substances*, EU, Luxembourg
12. Edwards *v.* National Coal Board (1949) IKB 704; (1949) 1 All ER 743
13. Health and Safety Executive, Environmental Hygiene Series Guidance Note No. EH 40, *Occupational Exposure Limits*, HSE Books, Sudbury, updated annually
14. Health & Safety Executive, *Investigation Report: Flixborough Disaster*, HSE Books, Sudbury (1975)
15. Health and Safety Commission, *Consultative Document on the Control of Major Accident Hazards Regulations*, HSE Books, Sudbury (1998)
16. Health & Safety Executive, Health and Safety Regulation Booklet No. HS(R)16, *Guide to the Notifications of Installations Handling Hazardous Substances Regulations 1982*, The Stationery Office, London (1983)
17. European Union, Directive No. 82/501/EEC, *Council Directive on Major Accident Hazards of Certain Industrial Activities*, EU, Luxembourg (1982)
18. Health & Safety Executive, Health and Safety Regulation Booklet No. HS(R)21, *Guide to the Control of Industrial Major Accident Hazards Regulations 1984*, HSE Books, Sudbury (1984)
19. Health & Safety Executive, Health and Safety Guidance Booklet No. HS(G)25, *Control of Industrial Major Accident Hazards Regulations 1984: Further guidance on emergency plans*, HSE Books, Sudbury (1985)
20. Health and Safety Commission, Legal Series Booklet No. L90, *Approved Carriage List, Information approved for the carriage of dangerous goods by road and rail other than explosives and radioactive material*, HSE Books, Sudbury (1996)
21. Health & Safety Executive, Legal Series Booklet No. L63, *Approved Guide to the Classification & Labelling of Substances Dangerous for Supply*, HSE Books, Sudbury (1995)
22. Health & Safety Executive, Health and Safety Guidance Series Booklet No. HS(G) 97, *A step by step guide to COSHH assessment*, HSE Books, Sudbury (1993)
23. Health and Safety Executive, Legal Series Booklet No. L5, *General COSHH ACOP and Carcinogens ACOP and Biological Agents ACOP* (1996 edn), HSE Books, Sudbury (1997)
24. Health & Safety Executive, Legal Series Booklet No. L86, *Control of Substances Hazardous to Health in Fumigation Operations: Approved Code of Practice: COSHH '94*, HSE Books, Sudbury (1996)
25. *The Control of Pollution Act 1974* (as amended by the Environmental Protection Act 1990), The Stationery Office, London (1990)
26. *The Special Waste Regulations 1996*, The Stationery Office, London (1996)
27. *The Control of Pollution (Supply and Use of Injurious Substances) Regulations 1986*, The Stationery Office, London (1986)

28. Health & Safety Executive, Health and Safety Series Guidance Booklet No. HS(G)40, *Chlorine from drums and cylinders*, HSE Books, Sudbury (1987)
29. Health & Safety Executive, Chemical Series Guidance Note No. CS4, *Keeping of LPG in cylinders and similar containers*, HSE Books, Sudbury (1986)
30. Health & Safety Executive, Health & Safety Guidance Series Booklets Nos HS(G)50, *The Storage of Flammable Liquids in Fixed Tanks (up to 10,000 m^3 total capacity)* (1990); HS(G)51, *The Storage of Flammable Liquids in Containers* (1990); HS(G)52, *The Storage of Flammable Liquids in Fixed Tanks (exceeding 10,000 m^3 total capacity)* (1991); HSE Books, Sudbury
31. British Distributors' & Traders' Association, *Warehousing of Chemicals Guide*, British Distributors' & Traders' Association, London (1988)
32. Health & Safety Executive, Health and Safety Guidance Series Booklet No. HS(G)71, *Storage of Packaged Dangerous Substances*, HSE Books, Sudbury (1992)
33. Health and Safety Commission, Consultative Document No. CD120, *Proposals for new petrol legislation*, HSE Books, Sudbury
34. Health & Safety Executive, Investigation Report (not numbered), *Fire and explosions at B & R Hauliers, Salford, 25 September 1982*, HSE Books, Sudbury (1983) (ISBN 0 11 883702 8)
35. Health & Safety Executive, Investigation Report (not numbered), *Fire and explosions at Cory's Warehouse, Toller Road, Ipswich, 14 October 1982*, HSE Books, Sudbury (1984) (ISBN 0 11 883785 0)
36. British Oxygen Company Ltd, *Safe Under Pressure, Guidelines for all who use BOC Gases in Cylinders*, British Oxygen Company, Guildford, Surrey (1993)
37. Health & Safety Executive, Legal Series Booklets Nos L89, *Approved Vehicle Requirements* (1996); L91, *Suitability of vehicles and containers and limits on quantities for the carriage of explosives: Carriage of Explosives by Road Regulations 1996–Approve Code of Practice* (1996); L92, *Approved requirements for the construction of vehicles for the carriage of explosives by road* (1996); L93, *Approved Tank Requirements: the provisions for bottom loading and vapour recovery systems of mobile containers carrying petrol* (1996); HSE Books, Sudbury.
38. Health & Safety Executive, Health and Safety Regulations Booklet No. HS(R)13, *Guide to the Dangerous Substances (Conveyance by Road in Road Tankers and Tank Containers) Regulations 1981*, HSE Books, Sudbury (1981)
39. *The Carriage of Dangerous Goods by Road (Driver Training) Regulations 1996*, The Stationery Office Ltd, London (1996)
40. *The Carriage of Explosives by Road Regulations 1996*, The Stationery Office Ltd, London (1996)
41. Chemical Industries Association, *Hauliers Safety Audit*, Chemical Industries Association, London (1986)
42. Kletz, T., *HAZOP & HAZAN – Identifying and Assessing Process Industry Hazards*, The Institution of Chemical Engineers, Rugby (ISBN 0 85 295285 6)
43. Chemical Industries Association, *A Guide to Hazard and Operability Studies*, Chemical Industries Association, London (1992)
44. British Standards Institution, BS EN 1050, *Safety of Machinery – Principle for Risk Assessment*, BSI, London (1997)
45. British Standards Institution, BS IEC 1025, *Fault Tree Analysis*, BSI, London
46. British Standards Institution, BS IEC 1508, *Functional Safety Life Cycle Management*, BSI, London
47. Bretherick, L., *Hazards in the Chemical Laboratory*, 4th edn, The Royal Society of Chemistry, London (1986)
48. Weston, R., *Laboratory Safety Audits & Inspections*, Institute of Science & Technology, London (1982)
49. The Royal Society of Chemistry, *Safe Practices in Chemical Laboratories*, The Royal Society of Chemistry, London (1989) (ISBN 0 851 86309 4)
50. The Society of Industrial Emergency Services Officers, *Guide to Emergency Planning*, Paramount Publishing Ltd., Boreham Wood (1986)
51. Chemical Industries Association, *Be prepared for an emergency – Training & Exercises*, Chemical Industries Association, London (1992) (ISBN 0 900623 73 X)
52. Health and Safety Commission, Consultative Document on Proposals for the Implementation of Dangerous Goods Safety Adviser's Directive, HSE Books, Sudbury (1998)

Chapter 8

The environment – issues, law and management

J. E. Channing

8.1 Introduction

The environment has long been a concern of society. The earliest concerns arose when pollution occurred which affected the well-being of fellow citizens. Such was the case in 1257 when Eleanor, Queen of Henry III, was forced to leave Nottingham because of the effects of smoke on her asthma and in 1273 Edward I prohibited the burning of coal in London as the smoke was deemed 'prejudicial to health'.

Anxiety over the way in which human activity can pollute air or water continues today. Further concerns arise because the scale and nature of human activity may now be threatening the earth as a planet. For these reasons protection of the environment has assumed greater significance.

8.2 Major environmental issues

We live in a modern, highly industrialised world with its striving for higher standards of living and the demand for ever more equipment to make work easier and leisure more enjoyable. Inevitably, this results in the generation of vast quantities of by-products as waste – gaseous, liquid and solid. These wastes now pose a threat to the environment.

Damage to the environment can occur in different ways and on different scales. They can have local or global implications. Some of the issues are considered below.

8.2.1 Global warming

In the balanced state of nature, gases emitted in the course of naturally occurring events – such as volcanic eruptions, forest fires, decaying vegetation and by living beings – are absorbed by plants and trees, which through a process of photosynthesis generate complex materials necessary for their growth and return the by-product, oxygen, to the atmosphere. However, the vast amounts of gases poured into the atmosphere by modern industrial processes and transport are too great for the reducing forest areas to cope with and much remains unabsorbed. The effect of this is to create an envelope of gases round the planet.

The earth derives its warmth from the sun and radiates any excess heat back into space as infrared radiations. But the envelope of gases absorbs the infrared rays reflected from the earth's surface and acts as an insulant preventing the surplus heat from dissipating into the upper atmosphere. The process is similar to the effect of glass in a greenhouse which allows the sun's energy to enter but prevents some of the radiated heat from escaping – the so-called 'greenhouse effect' caused by 'greenhouse gases'. When it involves the whole planet, it causes the global temperatures to rise, i.e. causes global warming.

The global warming potential of greenhouse gases, using the effect of carbon dioxide as a base of 1, are listed in *Table 8.1*.

Table 8.1 Global warming potential of greenhouse gases

Gas	*Global warming potential*	*Source*
Carbon dioxide	1	Fossil fuels, deforestation, road vehicles, forest fires
Methane	10	Agriculture, natural decay of vegetation, landfill
Nitrous oxide	100	Fossil fuels, agriculture, road vehicles
Chlorofluorocarbons (CFCs)	3000–7000	Refrigerants, aerosol sprays, solvents

There is concern over global warming and the extent to which the climate is changing with the consequent melting ice caps and rising sea levels. The most obvious culprits, CFCs, are being phased out as a result of international agreements and many governments are taking action by means of taxation and transportation initiatives to reduce the use of oil and petrol and hence the emission of greenhouse gases.

8.2.2 Ozone depletion

The earth is surrounded by a thin layer of ozone gas which is 25 km from the earth's surface. It filters out some of the potentially harmful ultraviolet (UV) radiation from the sun. Recent evidence from satellite data suggests that this ozone layer is being depleted, especially over the North and South Poles. This can exacerbate the problem of global warming and the melting of the polar ice caps. *Table 8.2* lists the gases which interact with the ozone to deplete the layer using chlorofluorocarbon gases as a base of 1.

Table 8.2 Ozone depleting potential of various gases

Gas	*Ozone depleting potential*	*Source*
Chlorofluorocarbons (CFCs)	1	Refrigeration, aerosol sprays, solvents
Carbon tetrachloride	1.1	Manufacturing intermediates
Halons	3–10	Fire extinguishants
1,1,1 trichloroethane	0.1	Degreasing agent

Scientific data on the effect of reducing the ozone layer are far from complete. An increase in surface UV radiation is thought to increase eye damage and skin cancers as well as damage the wider environment. International governmental concern has resulted in action being taken through the Montreal Protocol (1987), revised at the Rio de Janiero meeting of 1992, and aimed at phasing out some of these chemicals.

8.2.3 Acid rain

Acid rain is the name given to rain which has absorbed some of the acidic gases formed when fossil fuels are burnt. The sulphur and nitrogen compounds the fuels contain, often referred to as SO_x and NO_x, are first oxidised then dissolve in rainwater or snow to form dilute sulphuric and nitric acids. These acidify the soil resulting in crop failure, the wastage of forests and damage to the fabric of buildings. In addition the increased acidity of the soil results in aluminium being leached from the soil into rivers, killing fish and other aquatic creatures. The source of the original emission can be hundreds of miles from where the acid rain is deposited.

8.2.4 Photochemical smog

Photochemical smog occurs when strong sunlight acts upon a mixture of nitrogen oxides released from vehicle emissions which combine with the oxygen in the air to form ozone. In the presence of hydrocarbons, again from vehicle emissions, a number of complex substances are formed which condense into minute droplets and create the characteristic haze. Health effects can include irritation of the eyes and breathing problems. This smog tends to be most intense in the afternoon when the sunlight is most intense. It frequently occurs where high levels of vehicle emissions combine with strong sunlight and little wind to disperse the gases, such as in Los Angeles and Athens.

8.2.5 Waste disposal

Modern society produces large quantities of waste material all of which have, at some point, been derived from the natural resources of Mother Earth. Considerable amounts of waste are disposed of in landfill sites where fermentation generates methane and makes the ground unsuitable for agriculture or for habitable use. Whilst regulations govern the construction and use of waste sites to prevent the mixing of incompatible materials and prevent leachates from permeating to groundwater, concerns over the long-term impacts remain. Some waste materials are incinerated but emissions from the furnaces give rise to local pollution concerns. Waste disposal is a strictly regulated activity and is becoming more expensive as governments seek to reduce the quantities generated and encourage recycling, thereby conserving resources.

8.2.6 Water pollution

Water is a vital ingredient for life and of aesthetic importance in the enjoyment of leisure by many. Pollution of water destroys acquatic life and interferes with the riparian environment. The quality of surface waters, such as lakes and streams, is strictly controlled through the licensing of liquid discharges to water courses to prevent contamination by hazardous pollutants from domestic, commercial and industrial sources. In addition measures are taken to protect groundwater in underground strata, called aquifers, from contamination. Strict control is exercised over the discharges into the public sewers and drains to ensure that only permitted quantities of biologically contaminated waste fluids, chemicals or oils enter them and thence pass to the sewage treatment works. Similarly, discharges from the sewage treatment works into water courses are strictly controlled and monitored. River and groundwater are often used for drinking purposes and increasingly stringent water quality standards are being imposed. The overall effect is to encourage the use of less water, the recycling of process water and where discharge is necessary, it is of water that is less polluted.

8.2.7 Contaminated land

Land contamination has become a major issue as a result of the increasing need for housing land as the population expands or changes its living style. Protection of existing green belt and public objections to any nearby housing developments mean that planners must consider making use of former industrial land (so-called 'brownsites'). Previous industrial areas, now defunct but which operated at a time when scant attention was paid to environmental issues, are being considered as development sites. This raises concerns amongst potential users and private buyers that their health may suffer as a result of the chemicals that were allowed to drain into the ground or were disposed of by indiscriminate burying.

8.2.8 Resource depletion

The continued expansion of the human population and the increasing quantity and variety of products demanded by consumers are putting a great strain on the earth's resources. Depletion of those resources is a matter of serious concern. There is emphasis on reducing waste and increasing the recovery of materials by such techniques as recycling. Governmental initiatives include encouraging the better use of resources, reducing energy consumption and introducing new legislation to increase the recovery of packaging waste.

8.2.9 Noise and nuisance

In any area of habitable existence, a level of noise occurs that is accepted as part of that existence. In the activities of industry, commerce and the community, levels of noise are generated that are accepted as a necessary part of that activity. This is generally referred to as *background noise*. Acceptable levels of background noise vary with the time of day – higher levels are acceptable during daylight (working) hours than at night. They also differ geographically, such as between industrial, residential and countryside areas.

However, complaints can arise when the overall level of noise significantly exceeds the background level or when a persistent or piercing frequency is emitted. This can happen when machine bearings run dry, with transformer hum or from the exhaust outlet of a ventilation or air-conditioning system. Other features that cause annoyance can be changes of frequency of noise level, such as when a machine starts, changes speed or stops.

The measurement of background and community noise is a complex matter and specialists' assistance may be required in this area.

8.3 Environmental regulation

The United Kingdom subscribes to and is bound by international agreements and protocols. Such agreements are voluntary and can be viewed as 'soft law'. However, once ratified by a signatory country they must become national laws. Because of the global nature of many environmental issues, international agreements play a significant role in the development of environmental controls. The Montreal Protocol of 1987 and amendments tabled at the Rio de Janiero conference of 1992 were agreed by the European Union (EU) as a whole. They are being implemented by an EU regulation[2] aimed at phasing out the use of substances that deplete the ozone layer, notably CFCs.

EU regulations have direct applicability and are binding on Member States in their entirety. Another EU regulation[3] covers waste shipments and it, together with its UK manifestation[4], are aimed at controlling the movement of designated waste materials between Member States. It also covers the movement of wastes between Member States and OECD cosignatories of the Basle Convention which established the requirements for the control of such movements. A central feature of these particular agreements is the definition of 'wastes'. The international agreements classify wastes into 'Red', 'Amber' and 'Green' wastes:

- Red wastes are the most hazardous and include articles, substances and other wastes contaminated with polychlorinated biphenyl (PCB), asbestos and lead anti-knock compound sludges.
- Amber wastes include antifreeze fluids, used blasting grit, waste alumina, arsenic waste, lead acid batteries, phenols and other substances considered to pose a degree of environmental risk.
- Green wastes are, naturally, more extensive and include the following categories:
  (a) metal and metal alloy wastes in metallic non-dispersible form
  (b) other metal-bearing wastes arising from melting, smelting and refining of metals
  (c) wastes from mining operations in non-dispersible form
  (d) solid plastic wastes
  (e) paper, paperboard and paper products waste
  (f) glass waste in non-dispersible form
  (g) ceramic wastes in non-dispersible form
  (h) textile wastes
  (i) rubber wastes
  (j) untreated cork and wood wastes
  (k) wastes arising from agro-food industries
  (l) wastes arising from tanning and fellmongering operations and leather use
  (m) other wastes such as broken concrete and spent activated carbon.

Classification is important and extensive discussion on the technical arguments of ecotoxicity and health and safety hazards can occur. Indeed discussion often takes place on whether a material is a waste at all because the by-products of one process can be the feedstock of another.

The major environmental legislation in the UK is the Environmental Protection Act 1990 (EPA) which is supplemented by additional Acts and regulations. Further Acts and regulations deal with eco-labelling, planning, packaging and conservation. Potential safety and environmental issues occur when hazardous substances are used on industrial sites or are transported (see Chapter 7).

The environmental legislation is complemented by decided cases extending over 100 years, generally arising from a civil action for damages under the general tort of negligence. These cases clarify the interpretation of the wording of the laws and also suggest ways in which the duties imposed should be met.

8.3.1 Environment Protection Act 1990 (EPA)

This Act replaced the Control of Pollution Act 1974 and is concerned with all aspects of pollution of the environmental and, while laying down general requirements for controlling pollution, gives to the Minister concerned powers to make regulations in respect of particular environmental matters. These powers have been exercised and some of the regulations are considered in the following sections.

A multi-pronged approach to pollution control has been taken in the Act which covers:

- *Air Pollution Control* (APC) dealing with discharges into the atmosphere and includes noise in the community but not levels of noise within a workplace. It is enforced by the local authorities.
- *Integrated Pollution Control* (IPC) which encompasses control of pollution of the atmosphere, water and the ground and is enforced by the Environmetal Agency (EA).
- *Disposal of waste* to controlled sites enforced by local and county authorities through waste regulatory bodies.
- *Other nuisance matters* affecting the local community are enforced by the local authority and include matters such as statutory nuisances, stray dogs, supermarket trolleys etc.

The Act defines:

- *Pollutants* as
  - solid wastes for discharge onto land
  - liquid wastes whether discharged onto land or into waterways
  - discharges into the atmosphere
  - noise within the community
- *Controlled waste* as
  - household waste
  - industrial waste
  - commercial waste

  } These are qualified in the Controlled Waste Regulations 1992 – see below
- *Special waste* as
  - controlled waste that is so dangerous that it requires special disposal procedures, i.e. it is dangerous to life, or liquids with a flash point of 21°C or less. Details of the procedure to be followed are given in the Special Waste Regulations 1996 – see below.

The Minister has prescribed, in the Environmental Protection (Prescribed Processes and Substances) Regulations 1991, processes that cause environmental pollution and require authorisation before they can be operated. It is an offence to operate any of those processes without authorisation. In giving authorisation, a requirement may be imposed to use the *best available techniques not entailing excessive costs* (BATNEEC) – the environmental equivalent of *so far as is reasonably practicable*. The enforcing authorities are required to keep registers of authorisations and to make the registers available for public inspection.

The enforcing authority inspectors are given similar powers to HSE inspectors and can issue *Enforcement Notices* and *Prohibition Notices* (similar to Improvement and Prohibition Notices under HSW, respectively). Appeal against these Notices is to the Secretary of State. In any prosecution for an alleged offence, it is a defence if the accused can prove he used BATNEEC. Organisations storing, treating or disposing of controlled waste on land must have a *waste management licence.*

In exercising its IPC enforcement responsibilities, the EA takes a holistic approach to pollution control and sets conditions commensurate with current pollution technology that are the best practicable option for the environment as a whole (known as 'BPEO' – 'Best Practicable Environmental Option'). Account is taken of the total impact on water, land and air pathways; the ability of each or all of the pathways to absorb the pollutant; and the principles of sustainable development, i.e. taking into account the scarcity now or in the future of a natural resource.

8.3.2 Environmental Protection (Duty of Care) Regulations 1991

These Regulations expand on s. 34 of the EPA and their purpose to prevent the unregulated transportation of waste by imposing duties on all persons or companies involved in handling waste. The scope is wide and includes those who import, produce, carry, treat, or dispose of waste including any intermediaries. The duties relate to 'controlled waste' (all general waste but excluding household waste) and require transfer notes and records similar to, but less specific than, those required for special waste. The *duty of care* concept requires each party in the waste generation and disposal chain to take all reasonable measures to ensure the safe handling and disposal by appropriately licensed operators. The objectives are:

- to prevent the escape of controlled waste into the environment
- to ensure the waste is adequately packaged and labelled
- to ensure that waste is only handled by authorised or registered persons
- a written description of the waste to accompany it
- to ensure that reasonable care is taken that the above requirements are followed so that the waste is not disposed of improperly, in breach of any licence, or in a manner likely to cause pollution or harm to health and the environment.

In practice companies who dispose of waste should be able to demonstrate they have complied by auditing the disposal chain, from carrier to transfer station to landfill or other deposit. These Regulations are supported by a Code of Practice[5] which gives practical advice on the steps necessary to achieve compliance.

8.3.3 Water Resources Act 1991

This Act gives powers that enable both water quality and the amount of water abstracted for use to be regulated. Those powers extend to the control of the discharge of foul water to controlled waters through the issue of discharge consents or permits. A permit will list the characteristics of foul water that will be permitted for discharges and place limits on temperature, volume, rate of discharge and concentrations of specific chemicals. Controlled waters are defined as inland waters such as lakes, ponds and rivers, and groundwater such as aquifers and boreholes.

8.3.4 Water Industry Act 1991

This consolidating Act relates to the authorisation of water and sewerage companies who provide water or sewerage services. Discharge levels for trade effluent are set by the local water company after having consulted local dischargers and taken into account the ability of the treatment works to handle the effluent. Since discharges from sewage treatment works are normally to controlled waters, the constituent substances of trade effluent are considered carefully to ensure that any trace amounts of toxic chemicals are identified. Charges are made by the local water company for effluent treatment services based upon discharge volume, chemical oxygen demand (COD, a measure of the oxygen needed to oxidise completely all the carbonaceous matter in the effluent stream) and the solids content. The formula for making charges means that high volume chemically rich effluents attract heavy fees. By this means a financial incentive is applied to waste generators to reduce the quantity and chemical composition of the effluent.

8.3.5 Controlled Waste Regulations 1992

These regulations clarify part II of EPA in the definition of types of waste which are classified according to source:

- Household waste – any waste generated in a household but excluding:
  - mineral or synthetic oils and greases
  - asbestos
  - chemical waste

  but allows these latter to be collected for a charge.
- Industrial waste – waste from construction and demolition
  - laboratory waste
  - poisonous waste including toxic waste from laundering, dry cleaning and mixing and selling paints, pesticides, petrol etc.
- Commercial waste – from office and showroom
  - hotels
  - clubs, societies, associations, etc.
  - government buildings
  - markets and fairs.

8.3.6 Clean Air Act 1993

This Act consolidates several previous pieces of legislation. It prohibits the emission of dark smoke (darker than shade 2 on the Ringelmann chart) from any chimney apart from at start-up and during stoking and requires abatement equipment to be fitted to new furnaces. It regulates the height of chimneys and requires the control of smoke, grit and dust by the 'best practicable means' approach. Local authorities are empowered to designate smoke control areas.

8.3.7 Noise and Statutory Nuisance Act 1993

Nuisance is defined[6] as:

> an activity or state of affairs that interferes with the use or enjoyment of land or rights over land (*private nuisance*) or with the health, safety, comfort or property of the public at large (*public nuisance*).

However, *statutory nuisance* is defined in s. 79 of EPA as circumstances that are 'prejudicial to health or nuisance' and include:

(a) the state of any premises
(b) smoke emitted from any premises
(c) fumes or gases emitted from any premises
(d) any dust, steam, smell or other effluvia arising on trade, industrial or business premises
(e) any accumulation or deposit
(f) any animal kept in such a place as to pose a risk
(g) noise emitted from a premises
(ga) noise from a vehicle or equipment in a street
(h) any other matter.

Local authorities have duties to inspect their area to establish whether a statutory nuisance exists. The scope is wide but the Regulations specifically mention nuisance from:

- loudspeakers in streets and roads and
- audible intruder alarms.

Some guidance on the rating of environmental noise is provided by British Standard BS 4142[7].

8.3.8 Waste Management Licensing Regulations 1994

Section 33 of the EPA prohibits the disposal of controlled waste unless in accordance with a waste management licence, while s. 74 requires the licence holder to be a 'fit and proper person' without specifiying any qualifications. These Regulations refer to the degree of technical competence required of a licence holder and recognise the Waste Management Industry Training and Advisory Board's Certificate of Technical Competence as indicating the necessary level of competence.

For a transitional period pre-existing experience and qualifications have been deemed acceptable under the Waste Management Licensing (Amendments etc.) Regulations 1995.

The Regulations reinforce prohibitions concerning the deposit, treatment or storage of controlled waste unless in accordance with the terms of a waste management licence. Those activities which do not require a licence are identified. Other administrative procedures included cover the information required to apply for a licence, data to be kept on public registers and penalties for offences.

8.3.9 Environment Act 1995

This Act amends considerable parts of EPA. It established the Environmental Agency (EA) and laid down its authority and responsibilities. It also focused on contaminated land which is defined as 'land in a condition which by reason of substances in, on, or under it, significant harm or water pollution is being, or likely to be, caused'. Local authorities are required to identify such land and publish redemption notices specifying what must be done to clean it up. Wherever possible the organisation causing the contamination is required to pay for the clean-up (the 'polluter pays' principle). Further duties are placed on local authorities to designate air quality management areas. There is a broad recognition of the need to develop a national waste strategy and producers are to encouraged to reuse, recover or recycle waste products and materials. Powers are given to the enforcing authorities to deal with cases of imminent danger from waste.

8.3.10 Special Waste Regulations 1996

These Regulations list the types of waste considered to be 'special' which include materials deemed to be flammable, toxic, ecotoxic, corrosive, irritant and carcinogenic. It revokes the earlier Control of Pollution (Special Waste) Regulations 1980 but retains the documentary procedures laid down for these wastes. This involves a five part notification consisting of:

- notice of intended dispatch to the agency of the receiving area
- carrier's receipt for the waste given to the consignor
- advice note to consignee when waste delivered to disposal site
- carrier's copy of receipt from consignee
- notice of delivery to agency of the receiving area.

Signatures are required at the various stages of the transportation. These Regulations do not require the agency for the dispatching area to be notified of the removal of the waste from its area.

Landfill sites are licensed to take special waste up to designated quantities and with prescribed mixtures of different special wastes. When special wastes reach a landfill site they must be dealt with strictly in accordance with the practices stated in the licence. For example, asbestos waste must be buried immediately under 2 metres of controlled or household waste to prevent fibre release. Any one landfill site, or an individual cell on a multicell landfill site, will have a capacity limitation for categories of special waste. Advice on specific aspects of the handling of special wastes is given in publications of the Department of Environment[8].

The overall objective of the plethora of environmental legislation is to address the issues necessary to reduce the extent to which the environment is being polluted. The legislation is piecemeal but the action required to develop a co-ordinated environmental management system is set out in the BS EN ISO 14000 series of standards.

8.4 Environmental management systems

The emergence of the environment as a major issue has prompted the need to develop systems to enable all its aspects to be properly managed. The reference point for these environmental systems is the BS EN ISO 14000 series of standards. Those currently published are listed in *Table 8.3*. The Chemical Industries Association has for some years operated a combined environmental/health and safety management system known as 'Responsible Care'[9]. ISO 14000, however, has wider application and has the benefit of international recognition.

More standards are in development covering topics such as environmental labelling and site assessments.

The foundations of the environmental management system are to be found in Clause 4 of ISO 14001 and are represented in *Figure 8.1*.

The initial step is to establish an environmental policy with key features including:

- a commitment to comply with relevant legislation
- a commitment to the control of pollution and the continuous improvement of performance

Table 8.3 Current published standards in the ISO 14000 series

Standard	*Title*
ISO 14001	Environment management systems – Specifications with guidance for use
ISO 14004	Environmental management systems – General guidelines on principles, systems and supporting techniques
ISO 14010	Guideline for environmental auditing – General principles
ISO 14011	Guidelines for environmental auditing – Audit procedures – Auditing of environmental management systems
ISO 14012	Guidelines for environmental auditing – Audit procedures – Qualification criteria for environmental auditors
ISO 14040	Environmental management – Life cycle assessment. Principles and framework

SCHEMATIC OF ISO 14001

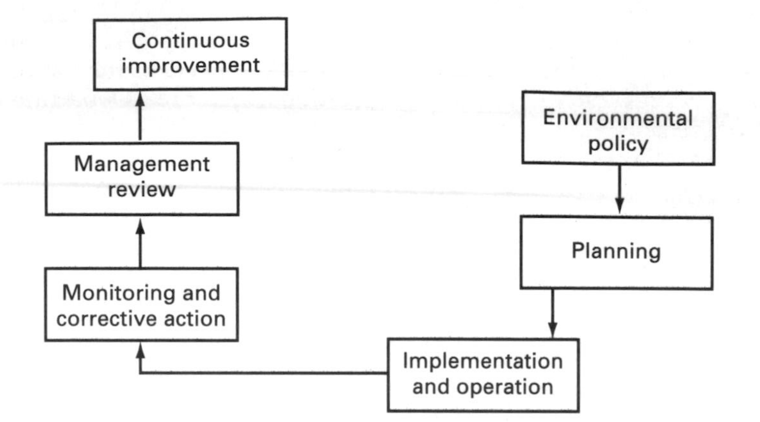

Figure 8.1 Diagram of a basic environmental management system

- arrangements to establish environmental objectives and targets
- a review process.

The policy should be documented, communicated to employees and available to the public. There is nothing in the standard which precludes the environmental policy being integrated with the health and safety policy.

Implementation of the policy must be planned. The first step is to recognise the environmental aspects of the organisation's activities which it can control or influence, and then to determine which of those activities have a significant impact on the environment. Whilst compliance with laws is an important component, the analysis of *aspects* and *impacts* should go beyond legal compliance. For example, a boiler in a power station emits gases (typically SO_x and NO_x) and as such falls within the environmental aspect of 'emissions to air' (Annex A of ISO 14001). It may be regulated by EPA and be in compliance with the authorisation to operate but the legally permitted discharge gases still impact on the environment.

Environmental management encompasses taking account of all the impacts arising from aspects such as:

- emissions to air
- releases to water
- waste management
- contamination of land
- use of raw materials and natural resources
- other local environmental and community issues.

The next step in the planning process is to choose from the list of impacts those that are to become objectives and which are consistent with the commitment to improve performance and prevent pollution. Objectives should be specific and accompanied by targets which are practicable, achievable and measurable.

Objectives may fall into three categories. An *Improvement Objective* is one which, for example, seeks a specific reduction in emissions or a stated increase in the proportion of waste that is recycled. A *Monitoring Objective* is one in which a particular impact is measured or closely monitored, for example by the continuous measurement of the amount of water being used or discharged. This latter type of objective may be a data gathering exercise which may be 'promoted', at a later date, to an Improvement Objective. The third type of objective is a *Management Objective* which is more general in nature and considers the arrangements in place to support the environmental system. An example could be 'to review induction training to ensure reference is made to environmental issues'. Each objective should have performance targets and completion dates with people assigned to undertake the relevant tasks.

The plan then needs to be implemented. The ISO standard expects that the organisation will have a defined structure with adequate resources where the roles and responsibilities of employees are stated and a manager appointed to oversee the implementation of the plan and to report on progress. There may be a need to provide training to ensure there is a degree of environmental awareness and that those participating are competent in the role they have to play. There should be proper communication of the plan throughout the organisation and an understood document control system. Arrangements should be in place to identify and control all operational aspects of the plan as they may affect the environment. This should include maintenance activities as well as the services provided by external suppliers. An emergency plan should be in place to cope with unintended incidents such as major spills, complaints from neighbours etc.

The standard requires that progress is monitored against the plan where the monitoring should include all aspects such as established objectives, targets and legal obligations. An internal audit should be carried out to identify non-conformances with the system and enable corrective or preventive actions to be initiated. In all this it is essential that good records are maintained.

The final key element is the periodic review of the entire system and plan by senior managers to ensure that the environmental improvements are being maintained and that the plan is effective.

The elements of an ISO 14000 system are similar to those of ISO 9000, the quality assurance standard, and hence allow the two systems to be integrated. Thus, one consistent management system can be developed that covers both quality assurance and environmental improvements.

8.5 Conclusion

The object of this chapter has been to provide a general overview of the scope and nature of environmental issues and how they can be managed. Managers and safety advisers need to have a working knowledge of environmental issues and how they interrelate with safety and health management issues. Specialist texts are available to enable a more detailed study of the subject and some are listed below.

References

1. *Environmental Protection. The Producer Responsibility (Packaging Waste) Regulations 1997*, The Stationery Office Ltd, London (1997)
2. European Union, Regulation No. 3093/94 *concerning substances that deplete the ozone layer*, EU, Luxembourg (1994)
3. European Union, Regulation No. 259/93/EEC *on the provision and control of shipments of waste within, into and out of the European Community*, EU, Luxembourg (1993)
4. *Environmental Protection, The Transfrontier Shipment of Hazardous Waste Regulations 1994*, The Stationery Office Ltd, London (1994)
5. *Waste Management. The Duty of Care, A Code of Practie*, The Stationery Office Ltd, London
6. *A Dictionary of Law*, 3rd edn, Oxford University Press, Oxford (1994)
7. British Standards Institution, BS 4142, *Method for rating industrial noise affecting fixed residential and industrial areas*, BSI, London
8. Department of the Environment, Transport and Regions, Waste Management Papers (WMP) series of advisory booklets:
   WMP 6 *Polychlorinated biphenyls*
   WMP 18 *Asbestos Waste*,
   The Stationery Office Ltd, London
9. Chemical Industries Association, *Responsible Care*, Chemical Industries Association, London

Other reading

Murley, L. (ed.), *Pollution Handbook*, National Society for Clean Air and Environmental Protection, 136 North Street, Brighton, East Sussex BN1 1RG (1994)

Sharratt, P. (ed.), *Environmental Management Systems*, Institution of Chemical Engineers, Davis Building, 165–189 Railway Terrace, Rugby, Warwickshire CV21 3HQ

Rothery, B., *ISO 14000 AND IS0 9000*, Gower Publishing Ltd, Gower House, Croft Road, Aldershot, Hampshire GU11 3HR

Sheldon, C. (ed.), *ISO 14001 and Beyond*, Greenleaf Publishing, Interleaf Productions Ltd, Broom Hall, Sheffield S10 2DR

Appendix 1

The Institution of Occupational Safety and Health

The Institution of Occupational Safety and Health (IOSH) is the leading professional body in the United Kingdom concerned with matters of workplace safety and health. Its growth in recent years reflects the increasing importance attached by employers to safety and health for all at work and for those affected by work activities. The Institution provides a focal point for practitioners in the setting of professional standards, their career development and for the exchange of technical experiences, opinions and views.

Increasingly employers are demanding a high level of professional competence in their safety and health advisers, calling for them to hold recognised qualifications and have a wide range of technical expertise. These are evidenced by Corporate Membership of the Institution for which proof of a satisfactory level of academic knowledge of the subject reinforced by a number of years of practical experience in the field is required.

Recognised academic qualifications are an accredited degree in occupational safety and health or the Diploma Part 2 in Occupational Safety and Health issued by the National Examination Board in Occupational Safety and Health (NEBOSH). For those assisting highly qualified OSH professionals, or dealing with routine matters in low risk sectors, a Technician Safety Practitioner (SP) qualification may be appropriate. For this, the NEBOSH Diploma Part 1 would be an appropriate qualification.

Further details of membership may be obtained from the Institution.

Appendix 2

Reading for Part I of the NEBOSH Diploma examination

The following is suggested as reading matter relevant to Part 1 of the NEBOSH Diploma examination. It should be complemented by other study.

Module 1A:	The management of risk	Chapters	2.1–all 2.2–paras. 8–11 2.3–all 2.4–paras. 1–3 3.8–paras. 1–6 **4.7–para. 11**
Module 1B:	Legal and organisational factors	Chapters	1.1–all 1.2–all 1.3–paras. 1–6 1.7–para. 2 1.8–all 2.2–paras. 13 and 14 2.6–paras. 1–4
Module 1C:	The workplace	Chapters	1.7–para. 2 3.6–all 3.7–all **4.2–all** **4.4–paras. 1–8** **4.6–paras. 2 and 4** **4.7–paras. 1, 2, 7 and 11**
Module 1D:	Work equipment	Chapters	**4.3–all** **4.4–all** **4.5–all**
Module 1E:	Agents	Chapters	3.1–all 3.2–all 3.3–all 3.5–paras. 1–6 3.6–all 3.8–paras. 4–7 **4.7–paras. 1–4**
Module 1CS:	Common skills	Chapter	2.5–para. 7

Additional information in summary form is available in *Health and Safety . . . in brief* by John Ridley published by Butterworth-Heinemann, Oxford (1998).

Appendix 3

List of abbreviations

ABI	Association of British Insurers
AC	Appeal Court
ac	Alternating current
ACAS	Advisory, Conciliation and Arbitration Service
ACGIH	American Conference of Governmental Industrial Hygienists
ACoP	Approved Code of Practice
ACTS	Advisory Committee on Toxic Substances
ADS	Approved dosimetry service
AFFF	Aqueous film forming foam
AIDS	Acquired immune deficiency syndrome
ALA	Amino laevulinic acid
All ER	All England Law Reports
APAU	Accident Prevention Advisory Unit
APC	Air pollution control
BATNEEC	Best available technique not entailing excessive costs
BLEVE	Boiling liquid expanding vapour explosion
BOD	Biological oxygen demand
BPEO	Best practicable environmental option
Bq	Becquerel
BS	British standard
BSE	Bovine spongiform encephalopathy
BSI	British Standards Institution
CBI	Confederation of British Industries
cd	Candela
CD	Consultative document
CDG	The Carriage of Dangerous Goods by Road Regulations 1996
CDG-CPL	The Carriage of Dangerous Goods by Road (Classification, Packaging and Labelling) and Use of Transportable Pressure Receptacle Regulations 1996
CDM	The Construction (Design and Management) Regulations 1994
CEC	Commission of the European Communities

CEN	European Committee for Standardization of mechanical items
CENELEC	European Committee for Standardisation of electrical items
CET	Corrected effective temperature
CFC	Chlorofluorocarbons
CHASE	Complete Health and Safety Evaluation
CHAZOP	Computerised hazard and operability study
CHIP 2	The Chemical (Hazard Information and Packaging for Supply) Regulations 1994
Ci	Curie
CIA	Chemical Industries Association
CIMAH	The Control of Industrial Major Accident Hazards Regulations 1984
CJD	Creutzfeldt–Jacob disease
COD	Chemical oxygen demand
COMAH	The Control of Major Accident Hazards Regulations (proposed)
COREPER	Committee of Permanent Representatives (to the EU)
COSHH	The Control of Substances Hazardous to Health Regulations 1994
CPA	Consumer Protection Act 1987
CTD	Cumulative trauma disorder
CTE	Centre tapped to earth (of 110 V electrical supply)
CWC	Chemical Weapons Convention

dB	Decibel
dBA	'A' weighted decibel
dc	Direct current
DETR	Department of the Environment, Transport and the Regions
DG	Directorate General
DNA	Deoxyribonucleic acid
DO	Dangerous occurrence
DSE(R)	The Health and Safety (Display Screen Equipment) Regulations 1992
DSS	Department of Social Services
DTI	Department of Trade and Industry

EA	Environmental Agency
EAT	Employment Appeals Tribunal
ECJ	European Courts of Justice
EC	European Community
EEA	European Economic Association
EEC	European Economic Community
EcoSoC	Economic and Social Committee
EHRR	European Human Rights Report

EINECS	European inventory of existing commercial chemical substances
ELF	Extremely low frequency
ELINCS	European list of notified chemical substances
EMAS	Employment Medical Advisory Service
EN	European normalised standard
EP	European Parliament
EPA	Environmental Protection Act 1990
ERA	Employment Rights Act 1996
ESR	Essential safety requirement
EU	European Union
eV	Electronvolt
EWA	The Electricity at Work Regulations 1989
FA	Factories Act 1961
FAFR	Fatal accident frequency rate
FMEA	Failure modes and effects analysis
FPA	Fire Precautions Act 1971
FSLCM	Functional safety life cycle management
FTA	Fault tree analysis
GEMS	Generic error modelling system
Gy	Gray
HAVS	Hand-arm vibration syndrome
HAZAN	Hazard analysis study
HAZCHEM	Hazardous chemical warning signs
HAZOP	Hazard and operability study
hfl	Highly flammable liquid
HIV+ve	Human immune deficiency virus positive
HL	House of Lords
HMIP	Her Majesty's Inspectorate of Pollution
HSC	The Health and Safety Commission
HSE	The Health and Safety Executive
HSI	Heat stress index
HSW	The Health and Safety at Work, etc. Act 1974
Hz	Hertz
IAC	Industry Advisory Committee
IBC	Intermediate bulk container
ICRP	International Commission on Radiological Protection
IEC	International Electrotechnical Committee (International electrical standards)
IEE	Institution of Electrical Engineers
IOSH	Institution of Occupational Safety and Health
IPC	Integrated polluton control
IQ	Intelligence quotient
IRLR	Industrial relations law report
ISO	International Standards Organisation
ISRS	International Safety Rating System

JHA	Job hazard analysis
JP	Justice of the Peace
JSA	Job Safety Analysis
KB	King's Bench
KISS	Keep it short and simple
LA	Local Authority
LEL	Lower explosive limit
$L_{EP,d}$	Daily personal noise exposure
LEV	Local exhaust ventilation
LJ	Lord Justice
LOLER	Lifting Operations and Lifting Equipment Regulations 1998
LPG	Liquefied petroleum gas
LR	Lifts Regulations 1997
lv/hv	Low volume high velocity (extract system)
mcb	Miniature circuit breaker
MEL	Maximum exposure limit
MHOR	The Manual Handling Operations Regulations 1992
MHSW	The Management of Health and Safety at Work Regulations 1992
MOSAR	Method organised for systematic analysis of risk
MPL	Maximum potential loss
M.R.	Master of the Rolls
NC	Noise criteria (curves)
NDT	Non-destructive testing
NEBOSH	National Examination Board in Occupational Safety and Health
NI	Northern Ireland Law Report
NIHH	The Notification of Installations Handling Hazardous Substances Regulations 1982
NIJB	Northern Ireland Judgements Bulletin (Bluebook)
NLJ	Northern Ireland Legal Journal
NONS	The Notification of New Substances Regulations 1993
npf	Nominal protection factor
NR	Noise rating (curves)
NRA	National Rivers Authority
NRPB	National Radiological Protection Board
NZLR	New Zealand Law Report
OJ	Official journal of the European Community
OECD	Organisation for Economic Development and Co-operation
OES	Occupational exposure standard
OFT	Office of Fair Trading
OR	Operational research

P4SR	Predicted 4 hour sweat rate
Pa	Pascal
PAT	Portable appliance tester
PC	Personal computer
PCB	Polychlorinated biphenyl
PHA	Preliminary hazard analysis
PMNL	Polymorphonuclear leukocyte
PPE	Personal protective equipment
ppm	Parts per million
ptfe	Polytetrafluoroethylene
PTW	Permit to work
PUWER	The Provision and Use of Work Equipment Regulations 1998
PVC	Polyvinyl chloride
QA	Quality assurance
QB	Queen's Bench
QMV	Qualifies majority voting
QUENSH	Quality, environment, safety and health management systems
r.	A clause or regulations of a Regulation
RAD	Reactive airways dysfunction
RCD	Residual cirrent device
RGN	Registered general nurse
RIDDOR	The Reporting of Injuries, Diseases and Dangerous Occurrences Regulations 1995
RM	Resident magistrate
RoSPA	Royal Society for the Prevention of Accidents
RPA	Radiation protection adviser
RPE	Respiratory protective equipment
RPS	Radiation protection supervisor
RR	Risk rating
RRP	Recommended retail price
RSI	Repetitive strain injury
s.	Clause or section of an Act
SAFed	Safety Assessment Federation
SC	Sessions case (in Scotland)
Sen	Sensitizer
SEN	State enrolled nurse
SIESO	Society of Industrial Emergency Services Officers
Sk	Skin (absorption of hazardous substances)
SLT	Scottish Law Times
SMSR	The Supply of Machinery (Safety) Regulations 1992
SPL	Sound pressure level
SRI	Sound reduction index
SRN	State registered nurse
SRSC	The Safety Representatives and Safety Committee Regulations 1977

SSP	Statutory sick pay
Sv	Sievert
SWL	Safe working load
SWORD	Surveillance of work related respiratory diseases
TLV	Threshold Limit Value
TUC	Trades Union Congress
TWA	Time Weighted Average
UEL	Upper explosive limit
UK	United Kingdom
UKAEA	United Kingdom Atomic Energy Authority
UKAS	United Kingdom Accreditation Service
v.	versus
VAT	Value added tax
VCM	Vinyl chloride monomer
vdt	Visual display terminal
VWF	Vibration white finger
WATCH	Working Group on the Assessment of Toxic Chemicals
WBGT	Wet bulb globe temperature
WDA	Waste Disposal Authority
WHSWR	The Workplace (Health, Safety and Welfare) Regulations 1992
WLL	Working load limit
WLR	Weekly Law Report
WRULD	Work related upper limb disorder
ZPP	Zinc protoporphyrin

Appendix 4

Organisations providing safety information

Institution of Occupational Safety and Health, The Grange, Highfield Drive, Wigston, Leicester LE18 1NN 0116 257 3100

National Examination Board in Occupation Safety and Health, NEBOSH, 5 Dominus Way, Meridian Business Park, Leicester LE3 2QW 0116 263 4700 Fax 0116 282 4000

Royal Society for the Prevention of Accidents, Edgbaston Park, 353 Bristol Road, Birmingham B5 7ST 0121 248 2222

British Standards Institution, 389 Chiswick High Road, London W4 4AL 0181 996 9000

Health and Safety Commission, Rose Court, 2 Southwark Bridge, London SE1 9HS 0171 717 6600

Health and Safety Executive, Enquiry Point, Magnum House, Stanley Precinct, Trinity Road, Bootle, Liverpool L20 3QY 0151 951 4000 or any local offices of the HSE

HSE Books, PO Box 1999, Sudbury, Suffolk CO10 6FS 01787 881165

Employment Medical Advisory Service, Daniel House, Trinity Road, Bootle, Liverpool L20 3TW 0151 951 4000

Institution of Fire Engineers, 148 New Walk, Leicester LE1 7QB 0116 255 3654

Medical Commission on Accident Prevention, 35–43 Lincolns Inn Fields, London WC2A 3PN 0171 242 3176

The Asbestos Information Centre Ltd, PO Box 69, Widnes, Cheshire WA8 9GW 0151 420 5866

Chemical Industry Association, King's Building, Smith Square, London SW1P 3JJ 0171 834 3399

Institute of Materials Handling, Cranfield Institute of Technology, Cranfield, Bedford MK43 0AL 01234 750662

National Institute for Occupational Safety and Health, 5600 Fishers Lane, Rockville, Maryland, 20852, USA

Noise Abatement Society, PO Box 518, Eynsford, Dartford, Kent DA4 0LL 01322 862789

Home Office, 50 Queen Anne's Gate, London SW1A 9AT 0171 273 4000
Fire Services Inspectorate, Horseferry House, Dean Ryle Street, London SW1P 2AW 0171 217 8728

Department of Trade and Industry: all Departments on 0171 215 5000
Consumer Safety Unit, General Product Safety: 1 Victoria Street, London SW1H 0ET
Gas and Electrical Appliances: 151 Buckingham Palace Road, London SW1W 9SS
Manufacturing Technology Division, 151 Buckingham Palace Road, London SW1W 9SS

Department of Transport
Road and Vehicle Safety Directorate, Great Minster House, 76 Marsham Street, London SW1P 4DR 0171 271 5000

Advisory, Conciliation and Arbitration Service (ACAS), Brandon House, 180 Borough High Street, London SE1 1LW 0171 396 5100

Health Education Authority, Hamilton House, Mabledon Place, London WC1 0171 383 3833

National Radiological Protection Board (NRPB), Harwell, Didcot, Oxfordshire OX11 0RQ 01235 831600

Northern Ireland Office
Health and Safety Inspectorate, 83 Ladas Drive, Belfast BT6 9FJ 01232 701444
Agricultural Inspectorate, Dundonald House, Upper Newtownwards Road, Belfast BT4 3SU 01232 65011 ext: 604
Employment Medical Advisory Service, Royston House, 34 Upper Queen Street, Belfast BT1 6FX 01232 233045 ext: 58

Commission of the European Communities, Information Office, 8 Storey's Gate, London SW1P 3AT 0171 222 8122

British Safety Council, National Safety Centre, Chancellor's Road, London W6 9RS 0171 741 1231/2371

Confederation of British Industry, Centre Point, 103 New Oxford Street, London WC1A 1DU 0171 379 7400

Safety Assessment Federation (SAFed), Nutmeg House, 60 Gainsford Street, Butler's Wharf, London SE1 2NY 0171 403 0987

Railway Inspectorate, Rose Court, 2 Southwark Bridge, London SE1 9HS 0171 717 6630

Inspectorate of Pollution, Romney House, 43 Marsham Street, London SW1P 3PY 0171 276 8083

Back Pain and Spinal Injuries Association, Brockley Hill, Stanmore, Middlesex 0181 954 0701

Appendix 5

List of Statutes, Regulations and Orders

Note: This list covers all four volumes of the Series. Entries and page numbers in bold are entries specific to this volume. The prefix number indicates the volume and the suffix number the page in that volume.

Appendix 6

List of Cases

Note: This list covers all four volumes of the Series. Entries and page numbers in bold are entries specific to this volume. The prefix number indicates the volume and the suffix number the page in that volume.

Appendix 7

Series Index

Note: This index covers all four volumes of the Series. Entries and page numbers in bold are entries specific to this volume. The prefix number indicates the volume and the suffix number the page in that volume.